FOOD AND BEVERAGE CONSUMPTION AND HEALTH

OCCURRENCES, STRUCTURE, BIOSYNTHESIS, AND HEALTH BENEFITS BASED ON THEIR EVIDENCES OF MEDICINAL PHYTOCHEMICALS IN VEGETABLES AND FRUITS

VOLUME 10

FOOD AND BEVERAGE CONSUMPTION AND HEALTH

Additional books and e-books in this series can be found on Nova's website under the Series tab.

BIOCHEMISTRY RESEARCH TRENDS

Additional books and e-books in this series can be found on Nova's website under the Series tab.

FOOD AND BEVERAGE CONSUMPTION AND HEALTH

OCCURRENCES, STRUCTURE, BIOSYNTHESIS, AND HEALTH BENEFITS BASED ON THEIR EVIDENCES OF MEDICINAL PHYTOCHEMICALS IN VEGETABLES AND FRUITS

VOLUME 10

NOBORU MOTOHASHI
EDITOR

Copyright © 2018 by Nova Science Publishers, Inc.

All rights reserved. No part of this book may be reproduced, stored in a retrieval system or transmitted in any form or by any means: electronic, electrostatic, magnetic, tape, mechanical photocopying, recording or otherwise without the written permission of the Publisher.

We have partnered with Copyright Clearance Center to make it easy for you to obtain permissions to reuse content from this publication. Simply navigate to this publication's page on Nova's website and locate the "Get Permission" button below the title description. This button is linked directly to the title's permission page on copyright.com. Alternatively, you can visit copyright.com and search by title, ISBN, or ISSN.

For further questions about using the service on copyright.com, please contact:
Copyright Clearance Center
Phone: +1-(978) 750-8400   Fax: +1-(978) 750-4470   E-mail: info@copyright.com.

NOTICE TO THE READER

The Publisher has taken reasonable care in the preparation of this book, but makes no expressed or implied warranty of any kind and assumes no responsibility for any errors or omissions. No liability is assumed for incidental or consequential damages in connection with or arising out of information contained in this book. The Publisher shall not be liable for any special, consequential, or exemplary damages resulting, in whole or in part, from the readers' use of, or reliance upon, this material. Any parts of this book based on government reports are so indicated and copyright is claimed for those parts to the extent applicable to compilations of such works.

Independent verification should be sought for any data, advice or recommendations contained in this book. In addition, no responsibility is assumed by the publisher for any injury and/or damage to persons or property arising from any methods, products, instructions, ideas or otherwise contained in this publication.

This publication is designed to provide accurate and authoritative information with regard to the subject matter covered herein. It is sold with the clear understanding that the Publisher is not engaged in rendering legal or any other professional services. If legal or any other expert assistance is required, the services of a competent person should be sought. FROM A DECLARATION OF PARTICIPANTS JOINTLY ADOPTED BY A COMMITTEE OF THE AMERICAN BAR ASSOCIATION AND A COMMITTEE OF PUBLISHERS.

Additional color graphics may be available in the e-book version of this book.

Library of Congress Cataloging-in-Publication Data

ISBN: 978-1-53614-141-2

Published by Nova Science Publishers, Inc. † New York

CONTENTS

Preface   vii

Chapter 1   Medicinal Phytochemicals and Health Effects of *Panax ginseng* (Korean Ginseng)   1
Noboru Motohashi, Anuradha Vanam, Jyothirmayi Vadapalli and Rao Gollapudi

Chapter 2   Coffee's Phytochemicals: From Biosynthesis to Health Benefits   93
Lourdes Valadez-Carmona, Carla Patricia Plazola-Jacinto, Marcela Hernández-Ortega and D. Nayelli Villalón-López

Chapter 3   Nature-Inspired Phytochemicals and the Pharmacological Activities of Herbal Plants of the Anacardiaceae Family and *Semecarpus anacardium* L.f.   133
Vustelamuri Padmavathi, Bhattiprolu Kesava Rao and Noboru Motohashi

Chapter 4	Carotenoids from Mexican Peppers and Their Beneficial Effects *Carla Patricia Plazola-Jacinto,* *Lourdes Valadez-Carmona,* *D. Nayelli Villalón-López* *and Marcela Hernández-Ortega*	**213**
About the Editor		**243**
Index		**245**

PREFACE

In Korea, China and Japan, and surroundings of them In East Asia, Korean ginseng has long been used as a private medicine for nourishing tonic, maintaining physical strength, various diseases treatment and prevention. Coffee is known for preventing lifestyle diseases such as high blood pressure, myocardial infarction, diabetes and dementia, and adjuvant effect. In India, *Semecarpus anacardium* L. f has been used as a cure for private medicine and has been used for various diseases treatment and prevention. In fact, *Semecarpus anacardium* L. f. contains bioactive ingredients such as antioxidant compounds. Mexican peppers with bright variegated colors are rich in plant pigment. Among these plant dyes, in particular, authors describe the correlation between carotenes and functionality, as well as the challenges to prevent and treat these diseases. Outline explanation of these four titles is made as follows:

Chapter 1 - *Panax ginseng* (Korean ginseng) has long been one of the representative traditional medicines mainly in the East Asian region including China, the Korean Peninsula, Taiwan, Ryukyu archipelago, Japan, *etc.* for prevention and treatment of many diseases, Additionally, *Panax ginseng* (Korean ginseng) medicinal efficacy has been recognized and has been used up to now despite being rare and expensive from long ago. However, the curative effect of *Panax ginseng* (Korean ginseng) known so far is enormous and varied. Therefore, this Chapter describes the representative and known medicinal ingredients contained in *Panax ginseng* (Korean ginseng), pharmacological actions and mechanisms by experiments and clinical experiments using human beings and animals. Further, in detail, this Chapter only describes in a limited way on myocardial relaxation, antiobesity effect, angiogenesis inhibitory action, hypoglycemic effects, antiviral effects, erectile dysfunction (E.D.) or impotence, improvement effects of ginseng saponins (ginsenosides), effect on exercise performance, immunomodulating activities, memory enhancing and neuroprotective effects, and dosage as well as avoidance for administration.

Based on the results of these *Panax ginseng* experiments, the further modifications of functional ingredients in *Panax ginseng* will be undertaken in the future for improving and treating depression, and dementia, hypertension, cancer, diabetes, and heart failure, and it is expected to become one of bibles for the new drug development against lifestyle diseases as prevention and therapeutic drugs of diseases including diabetes and heart failure, *etc.*

Chapter 2 - Coffee is an important beverage that could be obtained from two varieties of coffee beans (robusta and arabica). The brew of the roasted beans is worldwide consumed due to its good organoleptic qualities and is consumed worldwide. Besides of its appreciated sensorial characteristics, this beverage contains different phytochemicals with health related effect. For this reason the purpose of this chapter is to summarize the presence phytochemicals in this and the principal techniques to quantify them and the beneficial effects that these molecules could exert.

Chapter 3 - It mainly deals with the Importance of Herbal plants, their Biological Activities, Chemical constituents and their Pharmacological activities of *Semecarpus anacardium* L.f., (Sa). In our survey, we found that, the Anacardiaceae family has 83 genera and 860 species existing as trees, shrubs and vines. We have selected *Semecarpus anacardium* L.f., for its high medicinal value in Ayurveda and Siddha systems. Herbal drugs are contributing much to the human health especially in 21st Century. Natural medicine improves the inner immune system of the human body and no adverse effects could be observed. Hence, the herbal drug acts more effectively than the modern medicine. Now there is a great demand and stress from the people and various Governments including Pharmaceutical Industries for the isolation of active chemical ingredients from the natural herbal plant species and also for the development of novel chemical synthetic strategies to make their availability in large quantities to satisfy the immediate needs of the suffering people. 121 plant-derived drugs are produced commercially from less than 90 species of higher plants. It is estimated that 80% of antitumor and anti-infectious drugs were already in the market or under clinical trial are of natural origin obtained directly or indirectly, Prior to World War II, a series of natural products isolated from higher plants became clinical agents and a good number are still in use today. Quinine from cinchona bark, morphine (*Papavera somniferum*), codeine from the latex of the opium poppy, digoxin from digitalis leaves; atropine (*Atropa belladonna*) and hyoscine from species of the Solanaceae continue to be in clinical use. The antibiotic era dawned during and after World War II due to the antibacterial effects of a whole series of natural products isolated from species of Penicillium, Cephalosporium, and Streptomyces. In the post-war years, there were relatively few discoveries of new drugs from higher plants with the notable exception of reserpine from the *Rauwolfia serprntina* heralding the age of the tranquillizers and also vinblastine and vincristine from *Catharanthus roseus* which were effective in cancer chemotherapy. During recent years, the attention of the pharmaceutical industry has switched once more to the natural world and this may be illustrated by reference to three clinical drugs, taxol, etoposide and artemisinin. Taxol is obtained from the bark of the western yew (Pacific yew), *Taxus brevifolia*. Due to the presence of several anticancer drugs isolated from Anacardiaceae family, we have selected *Semecarpus anacardium* L.f., for its high medicinal value in Ayurveda and Siddha systems and isolated several active constituents. The most important of these are *Toxicodendron radicans* (poison ivy) and sap of the plants was toxic, mostly in the genus *Toxicodendron* that takes its name from the Greek "toxicos" meaning poisonous and "Dendron" meaning tree. This genus was formerly included in the genus *Rhus* (Sumac). Others include *Toxicodendron vernicifluum* (varnish tree) and *Toxicodendron vernix* (poison sumac). Brushing a leaf or any other part of the plant releases 3-*n*-pentadecycatechol, an irritating oil causing a red itch vesicular rash that appears 12-24 hours after exposure and lasts 4-5 days. Other toxic components of this family contain *Cotinus coggygria* (smoke tree), *Semecarpus anacardium* (marking nut tree) and *Smodinginum argutum* (African poison oak), *Schinus molle* (pepper tree), Cashew (*Anacardium occidentale*), *Mangifera indica* (mango). *Semecarpus anacardium* L.f. belongs to the family (Anacardiaceae) is a deciduous tree distributed in the forests of the Western Ghats of India. In the Indian system of medicine, the plant is well known as Bhallataka (Sanskrit) and commonly known as 'marking nut' (English) and Kaadugeru (Kannada). *Semecarpus anacardium* L.f. (Anacardiaceae) is reported to possess many medicinal properties. Trees: up to 25 m tall; young branches: terete, tomentose, watery latex present, which on drying it turns black. The black corrosive juice of the pericarp contains 90% of oxy acid anacardic acid and 10% of higher nonvolatile alcohol called cardol, also contains catechol and a mono-hydroxy phenol called as anacardol. The most significant components of the *Semecarpus anacardium* L.f. oil are phenolic compounds. On exposure to air, phenolic compounds get oxidized to quinones. The oxidation process can be prevented by keeping the oil under nitrogen. Many of the well-known properties of marking nut oils are easily explainable by the catechol half and lipoid-soluble C15 chain. During exposure to air, the catechol ring might be oxidized to an orthoquinone which might impart the dark color and also implies polymerization. The vesicant nature and the indelible pigmentation due to the rapid formation of the orthoquinonoid intermediate. The absorption of the oil by the skin is obviously due to the lipid-soluble C15 chain. In addition to this, chemical and phytochemical analysis of *Semecarpus anacardium* nuts revealed the presence of bhilawanol, anacardic acid, I-4I,II-3I,4I,I-5,II5,I7hexahydroxy [I-3,II-8]biflavanone, I-4I,II-4I,I-5,II-5,I-7,II-7-hexahydroxy[I-3,II-8]biflavonone (3I,8-binaringenin), I-4I,II-4I,I-7,II-7-tetrahexahydroxy[I-3,II-8] biflavonone (3I,8-billiquiritigenin), tetrahydrobusta flavonone, tetrahydroamentoflavonone, amentoflavone, semecarpuflavonone, galluflavonone, jeediflavonone, semecarpetine, anacarduflavonone, nallaflavonone, anacardoside, bhilwanol analogs, flavonoids and its pharmacological activities such as hypocholesterolemic activity, anti-inflammatory, immunomodulatory activity, antioxidant, adjuvant, antimicrobial activity, hypoglycemic and anti hyperglycemic, breast cancer, acetylcholinesterase inhibitory activity, acute and sub chronic toxicity study, antimutagenic effect, bioactivity, renal cortical necrosis, hair growth.

Chapter 4 - Color in food is mainly related to the presence of different pigments, among which, carotenoids are the molecules responsible for the yellow, orange and red color of many fruits and vegetables. Different kind of carotenoids could be found in peppers, and besides to its organoleptic importance, there are several studies that have described the beneficial effect of the carotenoids extracted from peppers finding that this bioactive compounds could act as important antioxidants protecting cells and tissues, mainly due to its anti-inflammatory effects. All the benefits exerted by pepper`s carotenoids suggests that these fruits are an important alternative not only to improve food taste and color, but also could help to improve humans health.

Noboru Motohashi, PhD
April 30, 2018

In: Occurrences, Structure, Biosynthesis … ISBN: 978-1-53614-141-2
Editor: Noboru Motohashi © 2018 Nova Science Publishers, Inc.

Chapter 1

MEDICINAL PHYTOCHEMICALS AND HEALTH EFFECTS OF *PANAX GINSENG* (KOREAN GINSENG)

Noboru Motohashi[1,*], *Anuradha Vanam*[2,†], *Jyothirmayi Vadapalli*[3,‡] *and Rao Gollapudi*[4,§]

[1]Meiji Pharmaceutical University, Kiyose-shi, Tokyo, Japan
[2]Sri Venkateswara University, Tirupathi, AP, India
[3]Acharya Nagarjuna University, Nagarjunanagar, AP, India
[4]University of Kansas, Lawrence, Kansas, US

ABSTRACT

Since ancient times, *Panax ginseng* (Korean ginseng) has been used traditionally as folk medicine in Korean Peninsula and surrounding area, mainly in China, Japan *etc*. *Panax ginseng* (Korean ginseng) is used to treat and prevent various diseases such as hyperactivity, hypertension, diabetes, various cancers, and pathogenicity. Recently, a wide variety of medicinal plant ingredients in *Panax ginseng* (Korean ginseng) have been

---

[*] Corresponding Author Email: noborumotohashi@jcom.home.ne.jp.
[†] agollapdr@gmail.com.
[‡] jyothi.galactica@gmail.com.
[§] gollapudirao@ku.edu.

identified, and pharmacological actions as well as mechanisms of these ingredients are being elucidated. Therefore, the rationale of this review is to describe these representative medicinal ingredients in *Panax ginseng* (Korean Ginseng), their disease curing effects and pharmacological actions based on experimental evidences. In order to emphasize the relationship between chemical structures and pharmacological actions, we included these chemical structures in the present review. Furthermore, we explained mechanisms and actions with expressions that are as simple and easy to understand as possible. The intention of this review is to inspire researchers for further investigations and explorations in use of *Panax ginseng* (Korean ginseng) to treat various aliments.

Keywords: *Panax ginseng* (Korean ginseng), antiobesity effect, Ginsenoside-Rb 2, sphingosine kinase-1, erectile dysfunction (E.D.) or impotence, lack- stamina, actoprotectors and adaptogens, ginseng saponins (ginsenosides), polysaccharides, neuroprotective effect, dosage, and avoidance for administration

CHEMICAL CONSTITUENTS

- panaquilon (panaxin. panacin. ginsenin. **1**)
- 3-(4,5-di-methylthiazol-2-yl)-2,5-diphenyltetrazolium bromide (MTT. **2**)
- ginsenoside-Rb2 ((3β,12β)-20-[(6-O-α-L-arabinopyranosyl-β-D-glucopyranosyl)oxy]-12-hydroxydammar-24-en-3-yl 2-O-β-D-glucopyranosyl-β-D-glucopyranoside. **3**)
- adriamycin (doxorubicin. **4**)
- ginsenoside compound K (CK. **5**)
- ginsenoside Rb1 (**6**)
- sphingosine 1-phosphate (S1P. C_{17}-sphingosine 1-phosphate. $C_{18}H_{38}NO_5P$. **7**)
- sphingosine ($C17$-sphingosine. 2-amino-4-octadecene-1,3-diol. $C_{18}H_{37}NO_2$. **8**)
- sphinganine ($C_{18}H_{39}NO_2$. **9**)
- palmitic acid (hexadecanoic aci. C16:0. **10**)

- stearic acid (octadecanoic acid. C18:0. **11**)
- arachidic acid (eicosanoic acid. C20:0. **12**)
- behenic acid (docosanoic acid. C22:0. **13**)
- nervonic acid (C24:1. ω-9. **14**)
- lignoceric acid (tetracosanoic acid. C24:0. **15**)
- GM6001 (**16**)
- rosiglitazone (Rosi. **17**)
- macelignan (Mace. **18**)
- aminoimidazole-4-carboxamide-1-β-D-ribofuranoside (AICAR. **19**)
- β-actin (**20**)
- thapsigargin (**21**)
- insulin (**22**)
- serotonin (5-hydroxytryptamine. 5-HT. **23**)
- trazodone (**24**)
- bemitil (2-ethylthiobenzimidazole hydrobromide. **25**)
- ethomersol (**26**)
- bromantane (*N*-(2-adamantil)-*N*-(*p*-bromophenyl)-amine. **27**)
- chlodantan (**28**)
- ademol (**29**)
- ginsenoside-Rg3 (**30**)
- ginsenoside-Rg2 (**31**)
- ginsenoside-Ro (**32**)
- γ-amino- butyric acid (GABA. **33**)
- histamine (His. **34**)
- bradykinin (BK. **35**)
- acetylcholine (Ach. **36**)
- nicotine (Nic. **37**)
- muscarine (**38**)
- ginsenoside Rg1 (**39**)
- lactate (**40**)
- glucose (**41**)

- taurine (**42**)
- creatine (**43**)
- ammonia (**44**)
- glycerol (glycerin. **45**)
- ginsenoside Rg$_3$ (**46**)
- ginsenoside Rg$_5$ (**47**)
- ginsenoside Rk$_1$ (**48**)
- glutamate (glutamic acid. **49**)
- *N*-methyl-D-aspartate (NMDA. **50**)
- scopolamine (hyoscine. **51**)
- 2,2-diphenyl-1-picrylhydrazyl (DPPH. **52**)
- ascorbic acid (**53**)
- phenelzine (**54**)
- warfarin (**55**)

ABBREVIATIONS

70% VO_{2max}	70% maximal oxygen uptakes
75% VO_{2max}	75% maximal oxygen uptake
ACS	acyl-CoA synthetase
AD	Alzheimer's disease
AMPK	AMP-activated protein kinase
AVS-penogram	audiovisual stimulation penogram
BI	browning intensity
CK	creatine kinase
CK´	ginsenoside compound K
CNS	central nervous system
CoA	coenzyme A
CPR	C-peptide immunoreactivity
CPT-1	carnitine palmitoyltransferase-1
Da	difference a
Db	difference b
DPP-4 inhibitor	dipeptidyl-peptidase 4 inhibitor

E.D.	erectile dysfunction
ELISA	enzyme-linked immunosorbent assay
ER	stress endoplasmic recticulum
EY	extraction yield
FFAs	free fatty acids
GLP-1	glucagon-like peptide 1
GLUT4	glucose transporter type 4
GMP	ginseng marc polysaccharide
GOT	glutamic oxaloacetic transaminase
GPP	general pharmacological properties
GS	ginseng root saponin
GS fraction	ginsenoside
H_2O_2	hydrogen peroxide
HbA_{1c}. HbA1c	blood glycosylated hemoglobin A_{1c}
HUVEC	human umbilical vein endothelial cells
IC50	fifty percent inhibitory concentration
I/G	insulin/glucagon
IIEF-5	International Index of Erectile Function
IKKβ	inhibitory kappa B kinase β
IR	insulin receptor
i.v.	intravenous
JNK	c-Jun NH_2-terminal kinase
KRG	Korean red ginseng
L	lightness value
LC/MS/MS	liquid chromatography/tandem mass spectrometry
LDH	lactic acid dehydrogenase; lactate dehydrogenase
LPL	lipoprotein lipase
LPS	lipopolysaccharide
MAOI	monoamine oxidase inhibitor
MMP	matrix metalloprotease
MMPs	metalloproteinases
MTT	3-(4,5-di-methylthiazol-2-yl)-2,5-diphenyltetrazolium bromide
NEFAs	free fatty acids; non-esterified fatty acids

NFκB	nuclear factor-kappa B
NIDDM	non-insulin-dependent diabetes mellitus
NK	natural killer
NO	nitrite
NT	neurotensin
OAD	anti-diabetic drug
PBS	phosphate-buffered saline
P.I.	phagocytic index
PIIINP	amino terminal propeptide
plasma TG	plasma triglycerides
PPAR-γ	peroxisome proliferators-activated receptors-gamma
RER	respiratory exchange ratio
RLE	rat lung endothelial
ROIs	reactive oxygen intermediates
RT-PCR	reverse transcription-polymerase chain reaction
SAL	sterile saline
SARI	serotonin antagonist and reuptake inhibitors
SD rats	Sprague-Dawley rats
SKI II	sphingosine kinase inhibitor II
S1P	sphingosine 1-phosphate
SPHK1	sphingosine kinase 1
STPG	STPG composite capsule formulation
SU drug	sulfonylurea drug
TC	total cholesterol
TNF-α	tumor necrosis factor-alpha
TZD	thiazolidine drug
VCO_2 levels	carbon dioxide production levels
VEGF	vascular endothelium growth factor
VO_2	oxygen consumption
VO_{2max}	maximal oxygen uptake
WG	wild ginseng

1. INTRODUCTION

1.1. Origin of *Panax Ginseng*

The English word "ginseng" is derived from Hokkien Chinese *jin-sim* meaning "person and plant root" referring to the roots, characteristic forked shape that resembles legs of a person. The botanical genus name *Panax* means all-healing in Greek. The etymology of *Panax*, scientific name of *Panax ginseng*, is expressed in a word from the representation of the panacea "Universalmittel" derived from goddess of treatment called Panacea and is expressed from various etiologies of ginseng. It seems to be the most effective restorative drug against sedimentary symptoms of general vigor due to weakened state of various organs [1]. Korean ginseng or Chinese ginseng - *Panax ginseng* C.A. Meyer, *Phenax notoginseng, Panax quinquefolium* var. Ginseng Regel et Maack., *Panax quinquefolium* var. coreense Sieb., or *Panax Schinseng* var. Coreense Nees. - (Photos 1 and 2) are perennial herbs belonging to Order: Apiales, Family: Araliaceae, Genus, which is native to Korean Peninsula or Manchuria of Northeast China.

On the other hand, ginseng, the native of North America Canada region known as *Panax quinquefolium (quinquefolius)* L., *Aralia quinquefolia* Planch. et. Decrie., or *Aureliana canadensis* Lafit - are completely different kinds from Chinese ginseng or Korean ginseng.

In the 1610s, Korean ginseng was first time imported into Europe by Dutch merchant. However, as the price of Korean ginseng was very expensive, it was rare to appear in the European market.

1.2. Root and Preparation Method of *Panax Ginseng*

The medicinal part of Korean ginseng is the root. The methods of processing *Panax ginseng* (Korean ginseng) are very few. The main preparation method is to dig roots that are about 4 years old after germination of plant around August and September to remove subroots after washing with water.

However, removed subroots are also used for medicinal purpose. After storing for a while, roots are boiled for 5 minutes, immediately transferred to cold water and washed. The boiling process starts with addition of inferior quality roots followed by high quality roots. The washed roots are dried under sunlight as well as fire power in order to dry root skin, followed by the removal of remaining thin subroots. These washed roots are dried again and stored in a box for several days.

In other methods, roots are collected quickly after the harvest without exposing to sunlight and drying, or waiting for several years for cultivation. Therefore, Korean ginseng has various names basing on its origin or its preparation method or prepared root part.

Photo 1. Total herbs of *Panax ginseng* (Korean ginseng). Photographed by Noboru Motohashi at Tokyo Metropolitan Medicinal Plant Garden, Tokyo, Japan. 7/13/2007 Fri.

1.3. The Therapeutic Application of *Panax Ginseng* in Traditional Chinese Medicine (TCM) and the Medicinal Value of *Panax Ginseng*

As for traditional Chinese medicinal herbs, Korean ginseng is indispensable as a medicine for treatment, and hence praised greatly as "miraculous medicine of divine herb for resuscitation." There is a Korean anecdote saying that one sells his own body to purchase ginseng in order to save his seriously ill family members.

Photo 2. Fruits of *Panax ginseng* (Korean ginseng). Photographed by Noboru Motohashi at Tokyo Metropolitan Medicinal Plant Garden, Tokyo, Japan. 7/13/2007 Fri.

Thus, it is evident that many people believed in the greatness of Korean ginseng to treat multiple ailments. The range of therapeutic applications of Korean ginseng is extremely wide and mostly adapted to all kinds of diseases.

2. PHYTOCHEMICALS, AND THEIR HEALTH EFFECTS OF *PANAX GINSENG*

2.1. Panax Ginseng

2.1.1. Myocardial Relaxation of Wild Ginseng

In 1854, Dr. Garriques identified yellow amorphous powder panaquilon (panacin. **1**) (Figure 1) called after the original plant name from a North American Canadian ginseng *Panax quinquefolius* (American ginseng). The term panaquilon (**1**) is derived from word which represents elements of the scientific Latin name *Panax quinquefolius*. Panaquilon (panacin. **1**) is an amorphous, sweet substance of the saponin class [2].

In 1962, the structure of panaquilon (panacin. **1**) was proposed by chemical decomposition investigations of panaquilon (panacin. **1**) [3, 4].

In 1905, Dr. Fujitani N of Department of Pharmacology, Kyoto Medical University, Kyoto, Japan confirmed the amorphous glycoside panaquilon (**1**) molecular weight $C_{32}H_{54}O_{13} \cdot 1/2H_2O$ from Korean ginseng.

First-in a study on the action of panaquilon (**1**) for cold blooded animals (frogs)-*first*, panaquilon (**1**) directly infringed heart muscle of gold wire frog (*Rana esculenta*), causing paralysis.

Second-on changes on muscle spasm conditions by panaquilon (**1**) using gold wire frog (*Rana esculenta*), panaquilon (**1**) attenuated or eliminated the spasmolytic properties by directly invading the muscle.

Third-on changes on muscle work status, panaquilon (**1**) suppressed the physiological function immediately as soon as panaquilon (**1**) acted on the muscle.

Second, on the action of panaquilon (**1**) for warm blooded animals (rabbit), *first*, panaquilon (**1**) also affected the heart of a cold-blooded animal (frog) directly by attacking its myocardium and paralyzing it as described above, panaquilon (**1**) also showed similar actions for warm-blooded animals (rabbits). When injecting a small amount of panaquilon (**1**) intravenously (*i.v.*) into a rabbit, panaquilon (**1**) showed a slight blood pressure decrease. In addition, when injected with a large amount of panaquilon (1), a remarkable blood pressure drop appeared. *Second*, blood pressure drop caused by panaquilon (1) was also found to be a paralysis of vasomotor center. Nevertheless, the blood pressure lowering effect of panaquilon (1) was not caused by excitation of its vagus nerve.

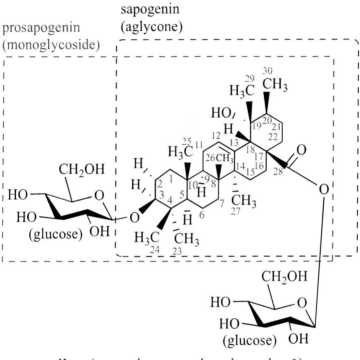

panaquilon (panaxin. panacin. ginsenin. 1)

Figure 1. Proposed structure of panaquilon (panaxin. panacin. ginsenin. 1).

From the above results, panaquilon (1) infringed frog's heart muscle and induced cardioplegia, while skeletal muscle of frog was also directly invaded by panaquilon (1), its spasmoly was reduced or eliminated. The skeletal muscle showed a remarkable reduction in total workload. The effect of panaquilon (1) on warm-blooded animals (rabbits) was relatively weak in general action. On the contrary the effect exerted on circulatory organs was remarkable. In frog's heart, panaquilon (1) invaded heart muscle itself, and showed remarkable hypotensive effect to attenuate heart's function. The reason for these activities is that panaquilon (1) attenuated as well as eliminated the physiological functions by invading myocardium and skeletal muscle [1, 3].

2.1.2. Antiobesity Effect of Wild Ginseng

Worldwide, regardless of age and sex, obesity is increasing which is one of the preventable diseases, occasionally resulting in death [5, 6]. The incidence rate of obesity in women is higher when compared to that (incidence rate of obesity) of men [5]. It is considered that obesity is one of the most serious public health problems in 21st century [7]. Number of obese patients in developed countries is more when compared to that (number of patients with obesity) in developing countries. Even in this era, obesity could be thought as a symbol of wealth as well as fertility and hence, the obesity onset tendency still remains in some parts of the world [6, 8].

Obesity is very closely related to incidence of type 2 diabetes [9]. The excessive weight gain by excess body fat in obese might frequently lead to development of insulin resistance.

Based on the facts above, in 2009, a group from College of Veterinary Medicine, Kyungpook National University, Daegu, Korea evaluated the antiobesity effect of wild ginseng (WG; *Panax ginseng* C.A. Meyer) by using male obese leptin-deficient (B6.V-Lepob, 'ob/ob') mice. For this purpose, antiobesity effect of wild ginseng was examined on body weight and blood glucose in nine-week-old male obese leptin-deficient (B6.V-Lepob, 'ob/ob') mice as an obesity animal model.

The whole plant of WG was washed, segmented, lyophilized powdered and taken with the food.

In this experiment, mice were divided into three groups – Group I mice (WG untreated control group), Group II mice (oral administration of WG 100 mg/kg body weight, once a day for 4 weeks), and Group III mice (oral administration of WG 200 mg/kg body weight, once a day for 4 weeks).

First, regarding the effect of WG oral administration during 4 weeks on body weight changes in ob/ob mice, body weight of WG-treated mice groups - Group II mice and Group III mice - showed dose-dependent

Medicinal Phytochemicals and Health Effects of Panax ginseng ... 13

decrease when compared to that (body weight changes) of Group I mice (WG untreated control group).

Second, regarding the effect of 4 weeks after WG oral administration on blood glucose level (mg/dL) in ob/ob mice, their blood glucose level (mg/dL) of two WG administration groups showed dose-dependent decrease like 140 mg/dL for Group II mice and 120 mg/dL for Group III mice, respectively, when compared to that (blood glucose level) of 160 mg/dL for Group I mice.

Third, *first*-regarding the effect of 4 weeks after WG oral administration on nuclear steroid receptor family with transcription regulatory function and essential transcription factor for adipocyte differentiation tissue peroxisome proliferators-activated receptors-gamma (PPAR-γ) mRNA expression in adipose tissue among three Groups (GI-GIII) in ob/ob mice, Group II and Group III received WG showed higher PPAR-γ mRNA expression (density) when compared to that (PPAR-γ mRNA expression) of Group I (control). However, there were no differences on PPAR-γ mRNA expression (density) among three Groups (GI-GIII) in skeletal muscle and liver in ob/ob mice.

Second-regarding the ffect of 4 weeks after WG oral administration on lipoprotein lipase (LPL) mRNA expression in adipose tissue among three Groups (GI-GIII), Group III showed the highest LPL mRNA expression (density: around 360), followed by around 250 for Group II, and around 75 for Group I (control).

Regarding the effect of 4 weeks after WG oral administration on LPL mRNA expression in skeletal muscle among three Groups (GI-GIII), Group II and Group III reduced their LPL mRNA expression (density) when compared to that (LPL mRNA expression density) of Group I (control).

Regarding the effect of 4 weeks after WG oral administration on LPL mRNA expression in liver among three Groups (GI-GIII), Group II and Group III slightly increased their LPL mRNA expression (density) when compared to that (LPL mRNA expression (density)) of Group I (control).

Fourth, it is generally known that skeletal muscle is the major site for insulin stimulated glucose disposal.

Regarding the effect of 4 weeks after WG oral administration a type 2 diabetes-related tissue glucose transporter type 4 (GLUT4) mRNA expression and insulin receptor (IR) mRNA expression (density) in adipose tissue, skeletal muscle, and liver in ob/ob mice among three Groups (GI-GIII), *first*, there were no differences on GLUT4 mRNA expression (density) among three Groups (GI-GIII) in adipose tissue in ob/ob mice.

Group II and Group III that received WG, significantly showed dose-dependently higher GLUT4 mRNA expression (density) when compared to that (GLUT4 mRNA expression (density)) of Group I (control) in skeletal muscle and liver in ob/ob mice. Especially, GLUT4 mRNA expression (density) in skeletal muscle showed the highest effect of WG.

Second, there were no differences on insulin receptor (IR) mRNA expression (density) among three Groups (GI-GIII) in adipose tissue in ob/ob mice.

Group II and Group III that received WG, significantly showed dose-dependently higher IR mRNA expression (density) when compared to that (IR mRNA expression (density)) of Group I (control) in skeletal muscle in ob/ob mice.

Group II that received WG, showed a decrease in the IR mRNA expression (density) when compared to that (IR mRNA expression (density)) of Group I (control) in liver in ob/ob mice. However, Group III that received WG showed higher IR mRNA expression (density) when compared to that (IR mRNA expression (density)) of either Group I (control) or Group II in liver in ob/ob mice.

Fifth, regarding the effect of 4 weeks after WG oral administration, histology study of adipose tissue (periepididymal adipose tissue) among three Groups (GI-GIII) in ob/ob mice, the adipocyte size of Group II and Group III that received WG, reduced in size when compared to that (adipocyte size) of Group I (control).

From the results above, it was shown that WG oral administration in mice significantly decreased body weight and blood glucose, and increased the PPAR-γ, GLUT4, LPL and IR mRNA expression level in tissues of ob/ob mice. These results suggested that WG can be one of the adjuvant therapies in metabolic syndromes, where abdominal circumference exceeded the standard visceral fat type obesity, a state with two or more of hyperglycemia, hypertension, hyperlipemia combined [10].

2.1.3. Antitumor Activity

In 2010, a group from Dept. of Food Science and Technology, Kyungpook National University, Daegu, Korea, evaluated the explosive puffing process for *Panax ginseng* C.A. Meyer roots, to compare the ginsenoside content and growth inhibitory effect using a dye used for cell death assay 3-(4,5-di-methylthiazol-2-yl)-2,5-diphenyltetrazolium bromide (MTT. **2**) (Figure 2) assay on white ginseng (dried ginseng), red ginseng (steamed and dried brown ginseng), explosively puffed ginseng of white ginseng (dried ginseng), red ginseng (steamed and dried brown ginseng), and three explosively puffed ginsengs (*PG1*, *PG2*, and *PG3*) from four-year-old dried roots of fresh raw *Panax ginseng* C.A. Meyer against four cancer cell lines – HT-29 cell lines of a human colorectal adenocarcinoma cell line with epithelial morphology, HeLa cells of cell lines derived from cervical cancer, MCF-7 cells of human breast cancer cell line, and HepG2 cells of human liver tumor cell line.

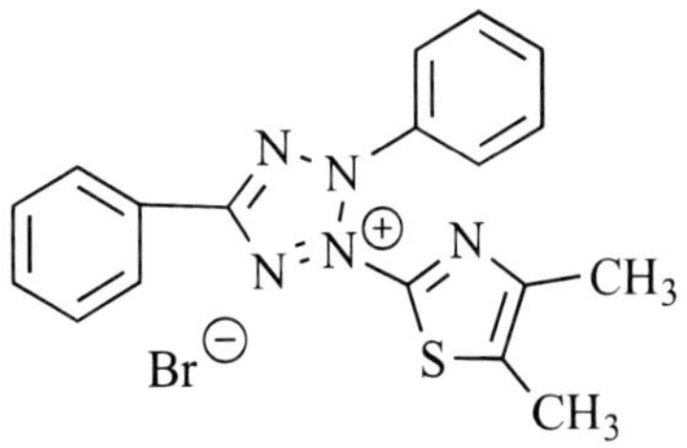

3-(4,5-di-methylthiazol-2-yl)-2,5-diphenyltetrazolium bromide (MTT. **2**)

Figure 2. 3-(4,5-di-Methylthiazol-2-yl)-2,5-diphenyltetrazolium bromide (MTT. 2).

First, on the comparison of extraction yield (EY) (%) and browning intensity (BI) at absorbance at 420 nm of white ginseng, red ginseng and explosively purified ginseng, where white ginseng prepared by drying at 60°C for 24 hrs (*WG*) were 54.69% (EY) and 0.108 (BI), followed by 53.83% (EY) and 0.724 (BI) for red ginseng prepared by steaming at 95°C for 2 hrs (*RG*), 53.93% (EY) and 0.284 (BI) for puffed ginseng prepared by puffing at 98 kPa (*PG1*), 55.87% (EY) and 0.440 (BI) for puffed ginseng prepared by puffing at 294 kPa (*PG2*), and 57.00% (EY) and 1.521 (BI) for puffed ginseng prepared by puffing at 490 kPa (*PG3*), respectively. From the results above, among 5 ginseng extractions, *PG3* showed the highest values - 57.00% (EY) and 1.521 (BI) - when compared to those (EYs and BIs) of other 4 ginseng extractions. *PG3* gave higher 57.00% (EY) when compared to that (EY) of either *WG* (EY: 54.69%) or *RG* (EY: 53.83%).

Second, on their comparison of Hunter color values of *WG*, *RG*, *PG1*, *PG2*, and *PG3* by using Hunter color difference meter, lightness value (*L*.), difference a (*Da*. component difference in reddish purple and blue-green direction), difference b (*Db*. component difference between yellow and blue direction) of *WG* was 82.91, 2.55, 17.71, followed by 75.70 (*L*), 4.33 (*Da*), 19.32 (*Db*) for *RG*, 73.19 (*L*), 7.14 (*Da*), 21.58 (*Db*) for *PG1*, 66.58 (*L*), 8.54 (*Da*), 21.09 (*Db*) for *PG2*, and, 59.37 (*L*), 8.73 (*Da*), 17.58 (*Db*) for *PG3*, respectively. [The abbreviations of *L*, *a*, *Da*, *b*, and *Db* represent as follows: lightness (*L*.) = lightness of sample; *L* − standard lightness (*a*)(indicating increase in redness) = difference *a* (*Da*); *L* − standard lightness (*b*) (indicating increase in yellowness) = difference *b* (*Db*)]. First, the *Da* values of three explosively puffed ginseng extractions (*PG1*, *PG2*, *PG3*) were higher when compared to that (*Da* value) of *RG*. The *Da* values of *PG2* (*Da*: 8.54) and *PG3* (*Da*: 8.73) were almost same values, however showed significant increase when compared to that (*Da*) of *PG1* (*Da*: 7.14).

Second, the *Db* values of *PG1* (*Db*: 21.58) and *PG2* (*Db*: 21.09) were almost same values, however, showed significant increase when compared to that (*Db*) of *PG3* (*Db*: 17.58). Additionally, the *Db* values of *WG* (*Db*: 17.71) and *PG3* (*Db*: 17.58) were almost same values, however, were lower when compared to that (*Db*) of *RG* (*Db*: 19.32).

Third, on the crude ginseng saponin contents (mg/g) in *WG*, *RG*, *PG1*, *PG2* and *PG3*, from 5 ginseng extractions, *RG* showed the highest around 75, followed by *PG3* (around 74), *PG2* (around 70), *PG1* (around 62), and *WG* (around 61), respectively. Hence, it is known that the general pharmacological properties (GPP) of *RG* were more active when compared to that (GPP) of *WG*. The steaming process of ginseng caused changes in

Medicinal Phytochemicals and Health Effects of Panax ginseng ... 17

the chemical compositions of ginseng saponin ginsenosides, thereby enhancing bioactivity of ginseng [11].

Fourth, *first*, on content comparison of eight ginsenosides (Re+Rg1, Rf, Rb1, Rc, Rb2, Rd, and Rg3), and total ginsenosides included in ginseng *WG*, *RG*, *PG1*, *PG2* and *PG3* from five ginseng extractions,

Re+Rg1 ginsenoside content (3.51 mg/g) in *WG* was the highest when compared to those (Re+Rg1 ginsenoside contents) (1.36-2.44 mg/g) of *RG*, *PG1*, *PG2* and *PG3*.

Rf ginsenoside content (1.36 mg/g) of *PG3* was the highest when compared to those (Rf ginsenoside contents: 0.52-1.09 mg/g) of *WG*, *RG*, *PG1*, and *PG2*.

Rb1 ginsenoside content (2.76 mg/g) of *WG* was the highest when compared to those (Rb1 ginsenoside contents: 1.42-2.35 mg/g) of *RG*, *PG1*, *PG2*, and *PG3*.

Rc ginsenoside content (1.06 mg/g) of *WG* was the highest when compared to those (Rc ginsenoside contents: 0.44-0.87 mg/g) of *RG*, *PG1*, *PG2*, and *PG3*.

Rb2 ginsenoside content (1.46 mg/g) of *RG* was the highest when compared to those (Rb2 ginsenoside contents: 0.55-1.19 mg/g) of *WG*, *PG1*, *PG2*, and *PG3*.

Rd ginsenoside content (0.39 mg/g) of *WG* was the highest when compared to those (Rd ginsenoside contents: 0.17-0.38 mg/g) of *RG*, *PG1*, *PG2*, and *PG3*.

First, Rg3 ginsenoside content (0.30 mg/g) of *RG* was the highest, followed by *PG3* (0.29), *PG2* (0.13), *PG1* (0.08), and *WG* (0.01).

Second, total ginsenoside content (10.08 mg/g) of *WG* was the highest, followed by *PG2* (7.82), *PG1* (7.30), *RG* (6.90), and *PG3* (5.77).

Fifth, regarding the comparison of antitumor activity (%) on crude saponin extract (crude saponin concentration: 1.0 mg/mL) from *WG*, *RG*, *PG1*, *PG2* and *PG3* against four cancer cell lines – HT-29 cell lines of a human colorectal adenocarcinoma cell line with epithelial morphology, HeLa cells of cell lines derived from cervical cancer, MCF-7 cells of human breast cancer cell line, and HepG2 cells of human liver tumor cell line, where 1st, antitumor activity (%) against HT-29 cell lines was the highest *PG1* (87.33), followed by *RG* (85.36), *PG2* (84.27), *PG3* (82.33), and *WG* (68.04), respectively. 2nd, antitumor activity (%) against HeLa cells was the highest *PG1* (84.88), followed by *PG3* (84.81), *RG* (84.52), *PG2* (83.82), and *WG* (62.00), respectively. 3rd, antitumor activity (%) against MCF-7 cells was the highest *RG* (69.10), followed by *PG3* (68.03), *PG2* (67.25), *PG1* (51.75), and *WG* (37.76), respectively. 4th, antitumor activity (%) against HepG2 cells was the highest *PG3* (78.69), followed by *RG* (73.98), *WG* (61.37), *PG2* (54.18), and *PG1* (44.61), respectively.

From the results above, according to the changes of 3 puffing pressures – 98 kPa (*PG1*), 294 kPa (*PG2*), and 490 kPa (*PG3*), the extraction yield increased, and the ginseng sample color showed significantly lower *L* (darker) and higher *a* (redder) values. Additionally, the crude saponon content (mg/g) in 3 puffed ginsengs (*PG1*, *PG2*, and *PG3*) was apparently higher when compared to that (crude saponon content) of *WG*.

Total ginsenoside content (5.77 mg/g) of *PG3* was the lowest when compared to those (total ginsenoside contents: 6.90-10.08 mg/g) of *WG*, *RG*, *PG1*, and *PG2*. However, Rg3 contents of *RG* (0.30 mg/g) and *PG3* (0.29 mg/g) were higher when compared to those (Rg3 contents) of *WG* (0.01 mg/g), *PG1* (0.08 mg/g), and *PG2* (0.13 mg/g).

Interestingly, the content of 5 ginsenosides (Re, Rg1, Rb1, Rc, and Rb2) in 3 explosively puffed ginsengs decreased, however, especially the content of ginsenoside Rg3 increased significantly according to changes of 3 different increasing puffing pressures, and reached the highest Rg3 value (0.29 mg/g) for *PG3* at a puffing pressure 490 kPa was higher when compared to that (Rg3 value) of *PG1* (0.08 mg/g), and *PG2* (0.13 mg/g). This Rg3 value (0.29 mg/g) of *PG3* was almost similar to Rg3 value (0.30 mg/g) of *RG*. Additionally, the increasing ginsenoside Rg3 content in 3 explosively puffed ginsengs showed increase in antitumor activities against four cancer cell lines – HT-29 cell lines, HeLa cells, MCF-7 cells, and HepG2 cells.

Therefore, in spite of lower total ginsenoside content of *RG* (0.30 mg/g) and *PG3* (0.29 mg/g), the antitumor activities (%) against 2 cancer cells (MCF-7 cells and HepG2 cells) of *RG* and *PG3* were higher when compared to those (antitumor activities) of *WG*, *PG1*, and *PG2*.

Here, it is known that *WG* changes the ginsenoside composition by heat- or steam-treatments [11, 12], and ginsenoside Rg3 is one of the main strong antitumor compound in ginsenosides [13, 14, 15].

Therefore, it suggests that antitumor activities of *RG* and *PG3* including other three samples (*WG*, *PG1*, *PG2*) were mainly caused by ginsenoside Rg3. At the same time, it also suggests that puffing process could provide one of materials for alternative means to produce functional *RG* products with additional advantage of reduced processing time [16].

2.1.4. Angiogenesis Inhibitory Action of Ginsenoside-Rb2

It is known that angiogenesis is a key process in promotion of cancer. Therefore, anti-angiogenic therapy has theoretic attraction: it may be less susceptible to development of treatment resistance because it is directed to stroma rather than to genomically unstable tumour cells [17].

In 1994, a group from College of Institute of Immunological Science, Hokkaido University, Sapporo, Japan examined the effect of a lipophilic constituent saponin ginsenoside-Rb2 (**3**) (Figure 3), derived from *Panax ginseng* roots on neovascularization and metastasis produced by highly metastatic subline murine B16-BL6 melanoma cells in specific pathogen-free female C57BL/6 mice and on the proliferation and invasion of endothelial cells *in vitro*.

First, on the effect of ginsenoside-Rb2 (**3**) on tumor-induced angiogenesis in mice, mice were intravenously (*i.v.*) administrated 3 different 10, 100, and 500 dose (μg/mouse) in the next day after injected dose (*i.d.*) inoculation of B16-BL6 cells (5 x 10^5) on the back. The number of vessels and tumor size (mm) of angiogenesis were around 13, around 3.8 for ginsenoside-Rb2 (**3**) 10 μg/mouse, around 9 and around 4.0 for ginsenoside-Rb2 (**3**) 100 μg/mouse, around 8, and around 4.0 for ginsenoside-Rb2 (**3**) 500 μg/mouse, respectively. These results suggested that administration of ginsenoside-Rb2 (**3**) dose-dependently prevented neovascularization, however, did not cause a significant decrease in tumor size.

Second, on the inhibition of tumor-induced angiogenesis by ginsenoside-Rb2 (**3**), the number of vessels and tumor size (mm) of angiogenesis were around 18 and. 5.6 for Ca^{2+}- and Mg^{2+}-free phosphate-buffered saline (PBS), and around 13 and 5.3 for ginsenoside-Rb2 (**3**) 100 μg/mouse *on Day 1*; around 25 and 6.4 for PBS, and around 18 and 5.3 for ginsenoside-Rb2 (**3**) 100 μg/mouse *on Day 3*; around 29 and 8.3 for PBS, and around 26 and 8.0 for ginsenoside-Rb2 (**3**) 100 μg/mouse *on Day 7*, respectively. On *Day 1*, *Day 3* or *Day 7* after tumor inoculation, the administration of ginsenoside-Rb2 (**3**) (100 μg/mouse) was effective in decreasing the number of vessels oriented toward tumor mass. Hence, the above results suggested that systematic administration of ginsenoside-Rb2 (**3**) was effective in the tumor-induced angiogenesis in mice.

Third, regarding the effect of administration route of ginsenoside-Rb2 (**3**) on tumor-induced angiogenesis, the number of vessels and tumor size (mm) of angiogenesis was around 15 and 4.8 for intravenous injection (*i.v.*) of PBS; around 9 and 4.1 for *i.v.* of ginsenoside-Rb2 (**3**) 100 μg/mouse; around 9 and 2.7 for intratumoral (*i.t.*) injection of ginsenoside-Rb2 (**3**) 100 μg/mouse; and around 9 and 3.2 for oral administration of ginsenoside-Rb2 (**3**) 2000 μg/mouse, respectively. These results suggested that all *i.t.*, *i.v.* and oral administration of ginsenoside-Rb2 (**3**) showed higher inhibition of angiogenesis when compared to that (inhibition) of PBS (a control). Additionally, *i.t.* and oral administrations showed higher inhibition of tumor growth at inoculated sites when compared to that (inhibition) of PBS (a control). These findings suggested that the treatment with ginsenoside-Rb2 (**3**) was effective in the prevention of new capillary formation at early stage of neovascularization.

Fourth, on the inhibition of invasion of rat lung endothelial (RLE) cells through Matrigel/Fibronectin-coated filters by ginsenoside-Rb2 (**3**), the invasion of capillary endothelial cells through surrounding tissues from a host vascular network toward tumor mass is one of the essential events of tumor neovascularization. The effect of ginsenoside-Rb2 (**2**) on endothelial cell invasion through a reconstituted basement membrane (Matrigel) was examined using a cell migration measurement device - Transwell chamber assay.

The number/field of invaded rat lung endothelial (RLE) cells was around 110 for control; around 24 for a positive control for 25 µg/mL adriamycin (4) (Figure 3); around.94 for 0.1 µg/mL ginsenoside-Rb2 (3), around 62 for 1 µg/mL ginsenoside-Rb2 (3), around 65 for 10 µg/mL ginsenoside-Rb2 (3), around 56 for 50 µg/mL ginsenoside-Rb2 (3), and around 51 for 100 µg/mL ginsenoside-Rb2 (3), respectively.

These results indicated that invasion of RLE cells through Matrigel to fibronection-coated lower surface of the filters was dose-dependently inhibited by ginsenoside-Rb2 (3). The preadministration of RLE cells with 25 µg/mL adriamycin (4) inhibited the invasion into Matrigel by around 80%.

Fifth, on the effect of 100 µg/mL ginsenoside-Rb2 (3) on experimental lung metastasis by intravenous injection (*i.v.*) of B16-BL6 melanoma cells, tumor-associated neovascularization is one of the important events in process of tumor metastasis. It was studied whether or not ginsenoside-Rb2 (3) could inhibit lung metastasis, produced by intravenous injection (*i.v.*) of B16-BL6 melanoma cells.

The intravenous (*i.v.*) treatment with various doses of 100 µg/mL ginsenoside-Rb2 (3) was begun on next day after tumor inoculation. The mice with B16-BL6 melanoma cells were killed 14 days after tumor inoculation, and lung tumor colonies were counted using a dissecting microscope.

The timing (on day) and number of tumor colonies (range) on day 14 was (- (untreated)), around 248 (205-283) for a control PBS; (1) around 185 (163-214) for 100 µg/mL ginsenoside-Rb2 (3); (1, 2, 3, 4), around 159 (133-200) for 100 µg/mL ginsenoside-Rb2 (3); (1, 2, 3, 4, 5, 6, 7, 8) and around 142 (95-172) for 100 µg/mL ginsenoside-Rb2 (3), respectively. These results revealed that both single administration (timing on 1 day only) and multiple administrations (4 timings and 8 timings) of 100 µg/mL ginsenoside-Rb2 (3) significantly showed inhibition of lung metastasis of B16-BL6 melanoma cells when compared to that (inhibition) of control PBS.

From the results 1 - V above, it is observed that ginsenoside-Rb2 (**3**) prevented angiogenesis and metastasis produced by mouse melanoma cells B16-BL6, and also inhibited invasion of endothelial cells *in vitro* [18].

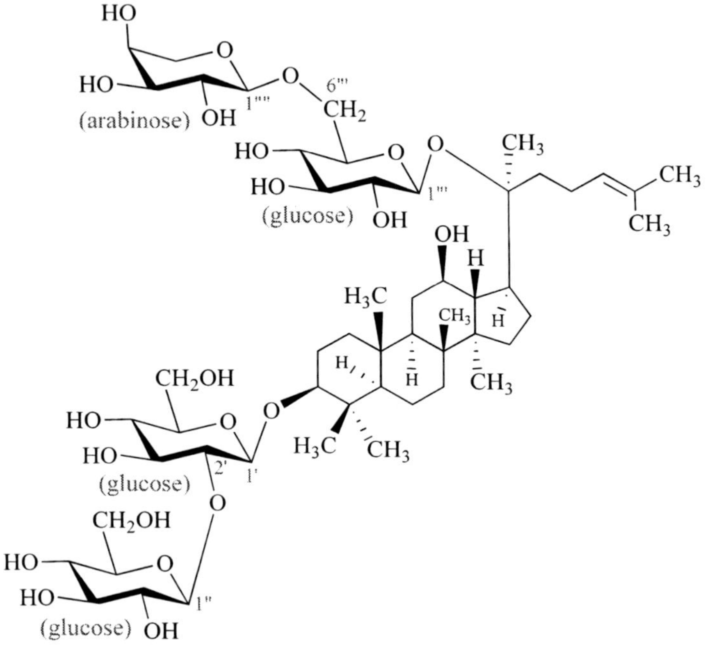

ginsenoside-Rb2 (**3**)

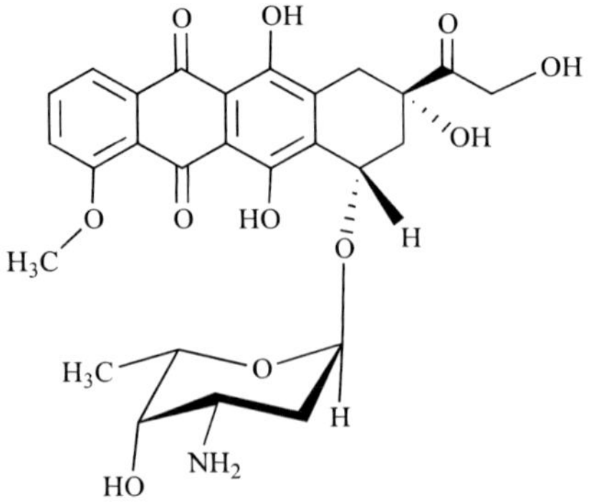

adriamycin (doxorubicin. **4**)

Figure 3. Ginsenoside-Rb2 (**3**), and adriamycin (doxorubicin. **4**).

2.1.5. Inhibition of Angiogenesis via Regulation of Sphingosine Kinase-1 by Ginsenoside Compound K (CK. 5)

It is known that both angiogenesis - formation of new blood vessels and proliferation of endothelium play important roles in metastasis and tumor growth, resulting in an increase in size of solid tumors. Angiogenesis inhibitors are considered to be effective as futuristic novel anticancer drugs. Vascular endothelium growth factor (VEGF) and its three receptors - VEGFR-1, VEGFR-2 and VEGFR-3 - constitute a key signaling system as drug targets that regulate proliferation and migration of vascular endothelial cells.

A ginseng saponin ginsenoside, compound K (CK. **5**) (Figure 4) - a major metabolite of ginsenoside Rb1 (**6**) (Figure 4) - from *Panax ginseng* root is generated by human intestinal bacterial flora, and rapidly absorbed from the gastrointestinal tract after oral administration, and slowly metabolized [19].

In 2011, a group from Department of Life and Nanopharmaceutical Sciences, Kyung Hee University, Seoul, Republic of Korea showed that ginsenoside compound K (CK. **5**) displayed inhibitory effect against basic fibroblast growth factor (FGF2)-induced neo-vascularization by Matrigel (Matrigel plugs assay) using cells excised from mice *in vivo* [20].

Moreover, in 2011, a group from Research and Development Center for Plant Medicine and Plant Drugs, Xiamen Overseas Chinese Subtropical, Plant Introduction Garden, Xiamen, China confirmed that nuclear export of nuclear factor (NF)-kappaB-p65, and reduction of metalloproteinases (matrix metalloproteinase 2 (MMP-2) and matrix metalloproteinase 9 (MMP-9) expression were associated with metastatic inhibition induced by compound K (CK) [21].

Based on the above facts, in 2014, a group from College of Pharmacy and MRC, Chungbuk National University, Cheongju, Korea examined the effect of ginsenoside compound K (CK. **5**) on sphingosine kinase 1 (SPHK1) activity and intercellular communication mediator sphingosine 1-phosphate (S1P. C_{17}-sphingosine 1-phosphate. **7**) (Figure 5) levels in human umbilical vein endothelial cells (HUVEC).

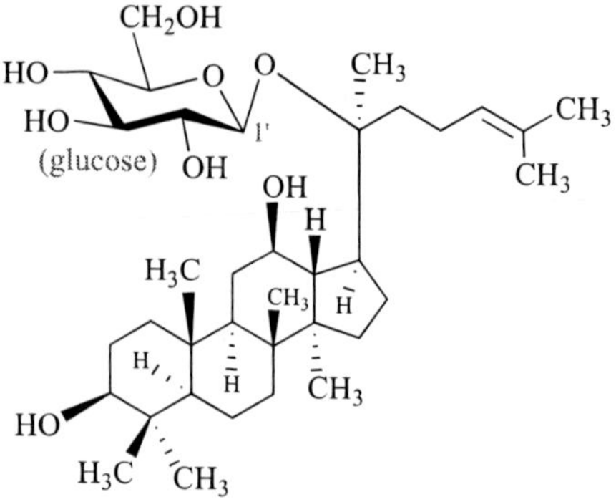

ginsenoside compound K (CK. **5**)

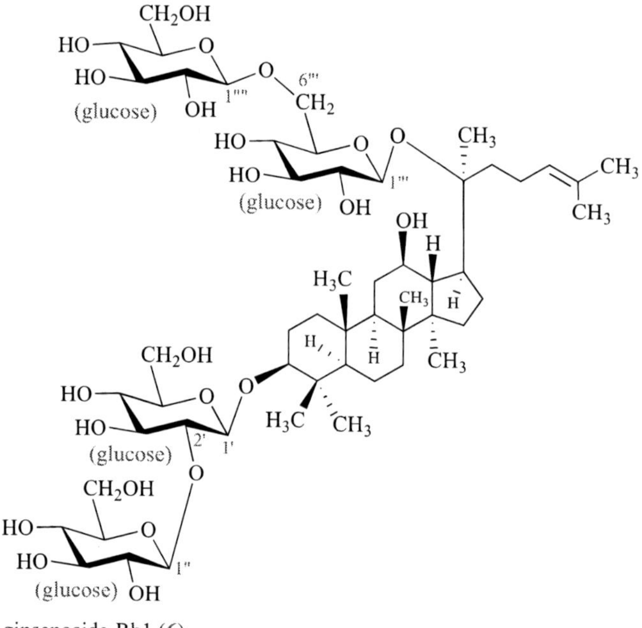

ginsenoside-Rb1 (**6**)

Figure 4. Ginsenoside compound K (CK. **5**) and ginsenoside-Rb1 (**6**).

First, on inhibitory effects of CK on HUVEC migration using an *in vitro* scratch wound-healing assay, migration cell density (fold increase) was around 10 for a control dimethyl sulfoxide (DMSO), around 90 for 100 nM sphingosine 1-phosphate (S1P. C_{17}-sphingosine 1-phosphate. **7**)-induced HUVEC migration, around 18 for 10 μM GM6001 (galardin) as a positive control (a broad spectrum matrix metalloprotease (MMP) inhibitor against 100 nM S1P (**7**), around 5.2 for 1 μg/mL ginsenoside compound K (CK. **5**) (CK-1) against 100 nM S1P (**7**), around 4.0 for 5 μg/mL ginsenoside compound K (CK. **5**) (CK-5) against 100 nM S1P (**7**), around 1.1 for 10 μg/mL ginsenoside compound K (CK. **5**) (CK-10) against 100 nM S1P (**7**), respectively. From these results, it was evident that 10 μg/mL ginsenoside compound K (CK. **5**) (CK-10) was apparently involved in the MMP enzymes in S1P (**7**)-induced HUVEC migration.

Second, in an investigation about the effects of ginsenoside compound K (CK. **5**) on activity of a physiologically active substance producing enzyme that causes cell migration sphingosine kinase 1 (SPHK1. sphingosine kinase), the effects of each with ginsenoside compound K (CK. **5**) (10 μg/mL in DMSO) on sphingosine 1-phosphate (S1P. **7**) synthesis in HUVEC after pretreatment with a natural substrate sphingosine (**8**) (Figure 5) (1 μM) was as follows: *first*-the concentration (pmol/mg protein) of sphingosine 1-phosphate (S1P. **7**) at 30 min before cell harvesting was around 10 for DMSO; around 200 for DMSO + 1 μM sphingosine (**8**); around 60 for a positive control 1 μM sphingosine kinase inhibitor II (SKI II) + 1 μM sphingosine (**8**); around 110 for ginsenoside compound K (CK. **5**) (10 μg/mL in DMSO) + 1 μM sphingosine (**8**); around 225 (Rd) - around 275 (Rh1) for each six ginsenoside compound (Rb1, Rc, Rd, Rg1, Rh1, and Ro) (10 μg/mL in DMSO) + 1 μM sphingosine (**8**); around 190 (Rf) - around 210 (F1) for each three ginsenoside (F1, Re, and Rf) (10 μg/mL in DMSO) + 1 μM sphingosine (**8**), respectively. From these results, it was observed that the pretreatment of HUVEC with 1 μM sphingosine (**8**) - a sphingosine kinase 1 (SPHK1) substrate - 30 min prior to cell collection increased the basal level of SPHK1 activity by 40-fold.

Nine ginsenosides (Rb1, Rc, Rd, Rg1, Rh1, Ro, F1, Re, and Rf) mildly activated cellular sphingosine kinase 1 (SPHK1) activity and produced a large amount of sphingosine 1-phosphate (S1P. **7**) from 1 μM sphingosine (**8**). Six ginsenoside compounds (Rb1, Rc, Rd, Rg1, Rh1, and Ro) significantly activated more sphingosine 1-phosphate (S1P. **7**) production by phosphorylating extracellularly added sphingosine (C17-sphingosine. **8**) when compared to that (sphingosine 1-phosphate (S1P. **7**) production) of a control DMSO. Two ginsenosides (Rb1 and Rh1) among nine ginsenosides (Rb1, Rc, Rd, Rg1, Rh1, Ro, F1, Re, and Rf) showed the most significant effect on sphingosine 1-phosphate (S1P. **7**) formation, indicating that structural specificity of ginsenosides may be required to activate sphingosine kinase 1 (SPHK1).

Second-on sphingosine 1-phosphate (S1P. **7**) activity which measured as the conversion rate of 2 nmol of sphingosine (C17-sphingosine. **8**) into C_{17}-sphingosine 1-phosphate (S1P. **7**) at 37 °C during a 30 min incubation, the C_{17}-sphingosine 1-phosphate (S1P. **7**) (pmol/mg/min) was around 0 for DMSO; around 1.08 for DMSO + 20 µM sphingosine (C17-sphingosine. **8**); around 0.44 for a positive control 1 µM SKI II + 20 µM sphingosine (C17-sphingosine. **8**); around 0.71 for ginsenoside compound K (CK. **5**) (10 µg/mL in DMSO) + 20 µM sphingosine (C17-sphingosine. **8**); around 1.19 (Rd) - around 1.31 (Rb1) for each six ginsenoside compound (Rb1, Rc, Rd, Rg1, Rh1, and Ro) (10 µg/mL in DMSO) + 20 µM sphingosine (C17-sphingosine. **8** around 1.11 (Re) - around 1.38 (F1) for each three ginsenoside (F1, Re, and Rf) (10 µg/mL in DMSO) + 20 µM sphingosine (C17-sphingosine. **8**), respectively.

These results suggested that the three ginsenosides (F1, Re, and Rf) did not show any effective inhibition of sphingosine 1-phosphate (S1P. **7**) synthesis. On the other hand, a hydrophobic ginsenoside compound K (CK. **5**) (around 0.71 for ginsenoside compound K (CK. **5**) (10 µg/mL in DMSO) + 20 µM sphingosine (C17-sphingosine. **8**)) inhibited the sphingosine 1-phosphate (S1P. **7**) production by sphingosine kinase 1 (SPHK1) by around 34% when compared to that (phingosine 1-phosphate (S1P. **7**) production) of the control (around 1.08 for DMSO + 20 µM sphingosine (C17-sphingosine. **8**)). This result is in agreement with observed inhibition of sphingosine 1-phosphate (S1P. **7**) production by ginsenoside compound K (CK. **5**), as shown in *first* result.

Third-on the inhibition of sphingosine kinase 1 (SPHK1) activity by ginsenoside compound K (CK. **5**) in dose-dependent manner, the C_{17}-sphingosine 1-phosphate (S1P. **7**) (pmol/mg/min) for sphingosine kinase 1

(SPHK1) activity was 0 for DMSO; around 1.08 for DMSO + 20 μM sphingosine (C17-sphingosine. **8**); around 0.44 for a positive control 1 μM SKI II + 20 μM sphingosine (C17-sphingosine. **8**); around 0.98 for ginsenoside compound K (CK. **5**) (2.5 μg/mL in DMSO) + 20 μM sphingosine (C17-sphingosine. **8**); around 0.81 for ginsenoside compound K (CK. **5**) (5 μg/mL in DMSO) + 20 μM sphingosine (C17-sphingosine. **8**); around 0.75 for ginsenoside compound K (CK. **5**) (10 μg/mL in DMSO) + 20 μM sphingosine (C17-sphingosine. **8**); around 0.60 for ginsenoside compound K (CK. **5**) (20 μg/mL in DMSO) + 20 μM sphingosine (C17-sphingosine. **8**), respectively. These results suggested that sphingosine kinase 1 (SPHK1) activity was dose-dependently inhibited by ginsenoside compound K (CK. **5**) from 2.5 to 20 μg/mL. Sphingosine kinase 1 (SPHK1) activity was not decreased any further at 40 μg/mL ginsenoside compound K (CK. **5**). Therefore, 20 μg/mL ginsenoside compound K (CK. **5**) displayed a maximal inhibitory effect on sphingosine kinase 1 (SPHK1) activity in this assay system.

Third, on the effects of four ginsenosides (Rb1, Rg1, CK (**5**), Rh1) for HUVEC viability using the absorbance of cells in culture medium containing WST-8 assay solution at 450 nm, the cell survival (%) was around 80 for 1 μg/mL ginsenoside compound K (CK. **5**); around 57 for 5 μg/mL ginsenoside compound K (CK. **5**); around 25 for 10 μg/mL ginsenoside compound K (CK. **5**), respectively. Ginsenoside compound K (CK. **5**) significantly reduced dose-dependently the HUVEC population. On the other hand, other three ginsenosides (Rb1, Rg1, Rh1) did not show the reduction of cell survival (%) by ginsenoside doses.

Fourth, on Western blot analysis of sphingosine kinase 1 (SPHK1) protein expression (42 kDa) in HUVEC treated with four ginsenosides (CK (**5**), Rg1, Rh1, Rb1) for 24 hrs, sphingosine kinase 1 (SPHK1) protein expression (fold increase) was around 1.0 for DMSO; around 0.9 for sphingosine kinase inhibitor II; around 0.7 for 5 μg/mL ginsenoside compound K (CK. **5**) (CK-5); around 0.6 for 10 μg/mL ginsenoside compound K (CK. **5**) (CK-10); around 1.1, 0.9, and 2.0 for each 10 μg/mL ginsenoside (Rg1, Rh1, and Rb1), respectively. These results illustrated that the treatment of HUVEC with 10 µg/mL ginsenoside compound K (CK. 5) (CK-10) for 24 hr reduced more significantly the expression (0.6) when compared to that (sphingosine kinase 1 (SPHK1) protein expression (fold increase): 0.9) of sphingosine kinase inhibitor II (SKI II). Especially, other three ginsenosides (Rg1, Rh1, and Rb1) showed no reduction of sphingosine kinase 1 (SPHK1) protein expression (fold increase) when compared to that (sphingosine kinase 1 (SPHK1) protein expression (fold increase) of either 5 µg/mL ginsenoside compound K (CK. 5) (CK-5) or 10 µg/mL ginsenoside compound K (CK. 5) (CK-10). It was concluded that the increase in sphingosine kinase 1 (SPHK1) activity and sphingosine 1-phosphate (S1P. C_{17}-sphingosine 1-phosphate. 7) production by these two ginsenosides (Rg1, Rh1) did not originate from increase in sphingosine kinase 1 (SPHK1) protein levels. Unexpectedly, treatment of HUVEC with ginsenoside (Rb1) increased the expression of sphingosine kinase 1 (SPHK1) protein twofold when compared to that (expression of the sphingosine kinase 1 (SPHK1) protein) of ginsenoside (Rh1). Ginsenoside (Rb1) is foremost discovered plant compound, inducing sphingosine kinase 1 (SPHK1) protein expression and sphingosine kinase 1 (SPHK1) activity.

Fifth, on the change of sphingolipid composition by DMSO, sphingosine kinase inhibitor II (SKI II), and ginsenoside compound K (CK. 5), the evaluation of the ability of ginsenoside compound K (CK. 5) to inhibit sphingosine kinase 1 (SPHK1) was also performed by analyzing sphingolipid metabolites.

First-the concentration (pmol/mg protein) of sphingolipid base) - both sphingosine (C17-sphingosine. **8**) and a saturated sphingosine constituting ceramide sphinganine (**9**) (Figure 5) - was around 17 for sphingosine (C17-sphingosine. **8**), around 9 for sphinganine (**9**) on DMSO (in DMSO); around 25 for sphingosine (C17-sphingosine. **8**), around 11 for sphinganine (**9**) on a positive control sphingosine kinase inhibitor II (SKI II) (1 µM in DMSO); around 37 for sphingosine (C17-sphingosine. **8**) and around 20 for sphinganine (**9**) on ginsenoside compound K (CK. 5) (10 µg in DMSO), respectively. These results disclosed that the concentrations of substrates - sphingosine (C17-sphingosine. **8**), and sphinganine (**9**) - in cell lysates by sphingosine kinase 1 (SPHK1) following treatment of HUVEC with ginsenoside compound K (CK. **5**) increased around twofold when compared to those (cell lysates) of DMSO.

Second, on the change of sphingosine 1-phosphate (S1P. C_{17}-sphingosine 1-phosphate. **7**) by DMSO, sphingosine kinase inhibitor II (SKI II), and ginsenoside compound K (CK. **5**), the concentration (pmol/mg protein) of sphingosine 1-phosphate (S1P. C_{17}-sphingosine 1-phosphate. **7**) was around 10.8 for DMSO (in DMSO); around 5 for positive control sphingosine kinase inhibitor II (SKI II) (1 μM in DMSO) an around 6.5 for ginsenoside compound K (CK. **5**) (10 μg in DMSO), respectively.

Third, on the change of sphingosine 1-phosphate (S1P C_{17}-sphingosine 1-phosphate **7**) by DMSO, sphingosine kinase inhibitor II (SKI II), and ginsenoside compound K (CK. **5**), the concentration (nM) of sphingosine 1-phosphate (S1P. C_{17}-sphingosine 1-phosphate. **7**) was around 0.35 for DMSO (in DMSO); around 0.18 for positive control sphingosine kinase inhibitor II (SKI II) (1 μM in DMSO) and around 0.28 for ginsenoside compound K (CK. **5**) (10 μg in DMSO), respectively.

These results suggested that the ginsenoside compound K (CK. **5**) (10 μg in DMSO) altered levels of sphingolipid metabolites by inhibiting sphingosine kinase 1 (SPHK1) activity. The concentration (pmol/mg protein) of pro-aptotic sphingoid bases (sphingosine (C17-sphingosine. **8**) and sphinganine (**9**)) was increased by approximately 2- to 3-fold following treatment of HUVEC with ginsenoside compound K (CK. **5**). In contrast, ginsenoside compound K (CK. **5**) inhibited sphingosine 1-phosphate (S1P. C_{17}-sphingosine 1-phosphate. **7**) synthesis and thus reduced extracellular secretion of sphingosine 1-phosphate (S1P. C_{17}-sphingosine 1-phosphate. **7**).

Sixth, on the concentrations of total ceramide species and individual ceramide species following treatment of HUVEC with ginsenoside compound K (CK. **5**) (10 μg in DMSO), *first*, the concentrations (p mol/mg protein)) of total ceramide was around 880 for DMSO (in DMSO); around 900 for positive control sphingosine kinase inhibitor II (SKI II) (1 μM in DMSO) and around 1320 for ginsenoside compound K (CK. **5**) (10 µg in DMSO), respectively. These results suggested that the treatment with ginsenoside compound K (CK. **5**) (10 µg in DMSO) resulted in 1.5-fold increase in amounts of total ceramide when compared to that (total ceramide species) of the control HUVEC treated with DMSO, indicating propagation of pro-apoptotic signals by total ceramide. As expected, sphingolipid rheostat (the ratio of ceramide/S1P) was also increased from 80.0 (basal) to 182.0 (SKI II) and to 212.9 (ginsenoside compound K (CK. **5**)), respectively.

Second, the concentrations (p mol/mg protein)) of six *N*-acyl chain lengthes for DMSO (in DMSO) on individual ceramide species was around 100 for C16:0 (**10**); around 10 for C18:0 (**11**); around 2 for C20:0 (**12**); around 2 for C22:0 (**13**); around 102 for C24:1 (**14**) and around 110 for C24:0 (**15**) (Figure 6), respectively.

The concentrations (pmol/mg protein)) of six *N*-acyl chain lengths for ginsenoside compound K (CK. **5**) (10 µg in DMSO) on individual ceramide species was around 135 for C16:0 (**10**); around 15 for C18:0 (**11**); around 2 for C20:0 (**12**); around 2 for C22:0 (**13**); around 120 for C24:1 (**14**); around 205 for C24:0 (**15**), respectively. These results demonstrated that the ceramide species in lipid extracts were further analyzed by liquid chromatography/tandem mass spectrometry (LC/MS/MS). A single ceramide with saturated fatty acid chain, tetracosanoic acid (C24:0. **15**), was significantly accumulated in HUVEC following ginsenoside compound K (CK. **5**) treatment.

Seventh, regarding the inhibitory effects of ginsenoside compound K (CK. **5**) on expression of two metalloproteinases (MMPs) - MMP2 and MMP9 using Western blot analysis, the fold increase (%) in metalloproteinase (MMP) expressions was MMP2 (0%) and MMP9 (0%) for DMSO; MMP2 (around 100%) and MMP9 (around 100%) for 100 nM sphingosine 1-phosphate (S1P. C_{17}-sphingosine 1-phosphate. **7**); MMP2 (around 14%) and MMP9 (around 22%) for a positive control 25 µM GM6001 (**16**) (Figure 6) + 100 nM sphingosine 1-phosphate (S1P. C_{17}-sphingosine 1-phosphate. **7**); MMP2 (around 55%) and MMP9 (around 82%) for 1 µg/mL ginsenoside compound K (CK. **5**) (CK-1) + 100 nM sphingosine 1-phosphate (S1P. **7**); MMP2 (around 45%) and MMP9 (around 80%) for 5 μg/mL ginsenoside compound K (CK. **5**) (CK-5) + 100 nM sphingosine 1-phosphate (S1P. **7**); MMP2 (around 18%) and MMP9 (around 50%) for 10 μg/mL ginsenoside compound K (CK. **5**) (CK-10) + 100 nM sphingosine 1-phosphate (S1P. **7**), respectively. These results suggested that ginsenoside compound K (CK. **5**) reduce dose-dependently both MMP2 expression and MMP9 expression in HUVEC. The cellular MMP expression was enhanced in HUVEC treated with 100 nM S1P (**7**), and the treatment of HUVEC for 24 hrs with ginsenoside compound K (CK. **5**) (10 μg/mL in DMSO) reduced significantly the expression of both MMP2 and MMP9 by around 80% and around 50%, respectively indicating that (First - Seventh), ginsenoside compound K (CK. **5**) inhibited sphingosine kinase 1 (SPHK1) activity and impaired collapse of homeostasis of sphingolipid metabolic balance by reducing the synthesis and extracellular release of sphingosine 1-phosphate (S1P. **7**). This activity resulted in the inhibition of HUVEC migration [22].

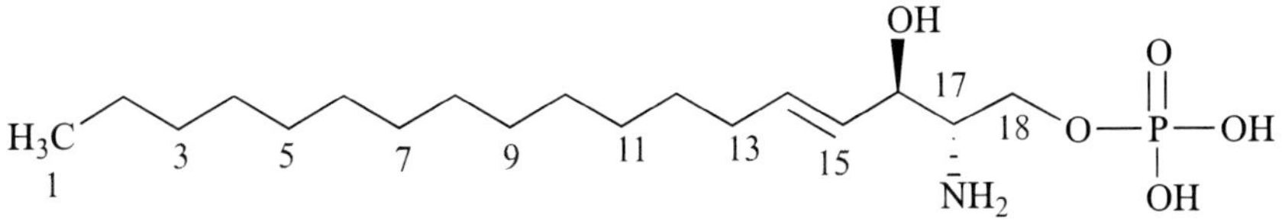

sphingosine 1-phosphate (S1P. C_{17}-sphingosine 1-phosphate. $C_{18}H_{38}NO_5P$. **7**)

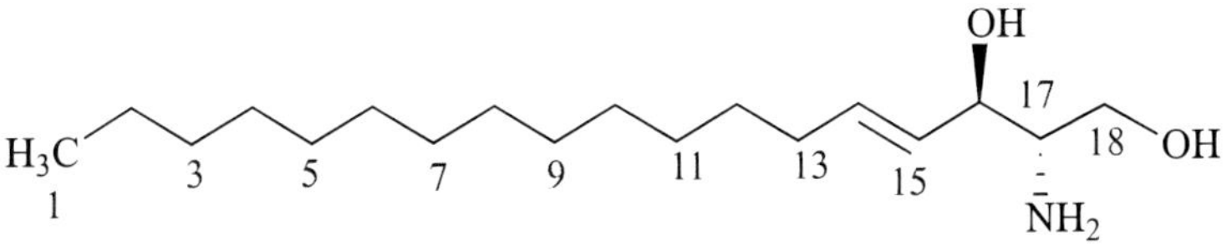

sphingosine (C17-sphingosine. 2-amino-4-octadecene-1,3-diol. $C_{18}H_{37}NO_2$. **8**)

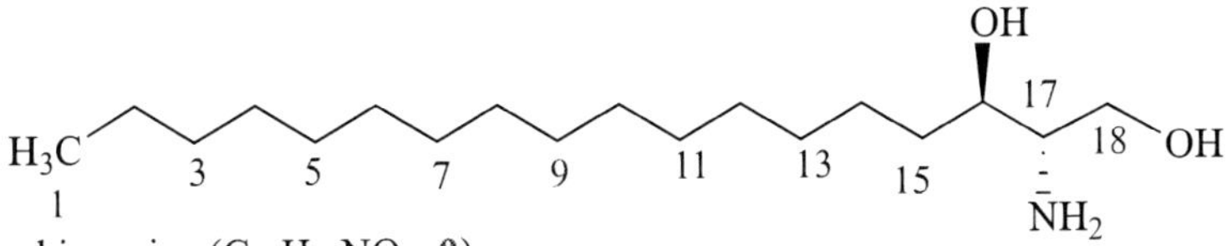

sphinganine ($C_{18}H_{39}NO_2$. **9**)

Figure 5. Sphingosine 1-phosphate (S1P. C_{17}-sphingosine 1-phosphate. $C_{18}H_{38}NO_5P$. **7**), sphingosine (C_{17}-sphingosine. 2-amino-4-octadecene-1,3-diol. $C_{18}H_{37}NO_2$.**8**), and sphinganine ($C_{18}H_{39}NO_2$. **9**).

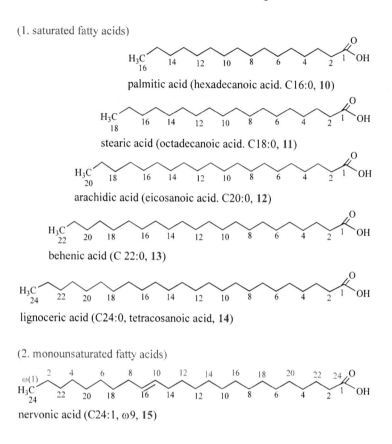

Figure 6. Palmitic acid (hexadecanoic acid. C16:0, **10**), stearic acid (octadecanoic acid. C18:0, **11**), arachidic acid (eicosanoic acid. C20:0, **12**), behenic acid (C 22:0, **13**), nervonic acid (C24:1, ω9, **14**), and lignoceric acid (C24:0, tetracosanoic acid, **15**) from *Panax ginseng*. GM6001 (**16**).

2.1.6. Hypoglycemic Effects

2.1.6.1. Effect on Type 2 Diabetes

The efficacy was evaluated with psychophysical tests and measurements of glucose balance, serum lipids, aminoterminal propeptide (PIIINP) concentration, and body weight for the hypoglycemic effects of *Panax ginseng* extract on type 2 diabetic patients with newly diagnosed non-insulin-dependent diabetes mellitus (NIDDM), after 36 NIDDM patients were treated for 8 weeks with *Panax ginseng* extract (100 or 200 mg) or placebo. The aminoterminal propeptide (PIIINP) is insulinotropic peptide derivative in which the amino-terminal amino acid is mutated. As a result, oral administration of *Panax ginseng* extract (100 or 200 mg) elevated mood, improved psychophysical performance, and reduced both fasting blood glucose (FBG) and body weight. The administration of 200 mg *Panax ginseng* extract improved significantly glycated hemoglobin, serum aminoterminal propeptide (PIIINP), and physical activity. Placebo reduced body weight and also altered serum lipid profile, however did not alter the fasting blood glucose (FBG). Hence, it was suggested that the oral administration of *Panax ginseng* extract was effective as adjunct in the treatment of type 2 diabetic patients [23].

Diabetes mellitus type 2 (type 2 diabetes) is characterized by both, the decrease of insulin secretion by pancreas and resistance against the action of insulin in different tissues such as muscle, liver and adipose, by leading to impaired glucose uptake which were caused by complex interactions of multiple factors [24]. Therefore, type 2 diabetes treatment is based on diet therapy and exercise therapy. Generally type 2 diabetes patients also need the oral administration of anti-diabetic drug (OAD) such as sulfonylurea drug (SU drug) or fast acting insulin secretagogue (non-sulfonylurea secretagogues), which acts on beta cells of pancreas, promotes insulin secretion for several hours, and lowers blood glucose level; biguanide drug, which suppresses sugar making in liver, promote use of glucose in muscle *etc.*, lower blood sugar level; α-glucosidase inhibitor, which delays decomposition/absorption of glucose in small intestine and suppresses the rise in blood glucose level rapidly after a meal.; thiazolidine drug (TZD), which improves effectiveness of insulin with fat and muscle, increase the use of glucose in blood, lower blood sugar; dipeptidyl-peptidase 4 inhibitor (DPP-4 inhibitor), which enhances the function of glucagon-like peptide 1 (GLP-1), a hormone that promotes secretion of insulin. GLP-1 is secreted from small intestine after food intake.

However, it is known that these anti-diabetic drugs (OADs) have adverse effects such as hypoglycemia, weight gain for sulfonylurea drug (SU drug), hypoglycemia for non-sulfonylurea secretagogues, hypoglycemia, gastrointestinal disorders, lactic acidosis for biguanide drug, increased stomach and fart, low blood sugar for α-glucosidase inhibitor, hypoglycemia, swelling, hepatic disorder, weight gain for thiazolidinedione (TZD), low blood sugar, gastrointestinal problems for dipeptidyl peptidase IV (DPP-4) inhibitor, respectively.

Based on the above facts, in 2011, a group from School of Korean Medicine, Pusan National University, Beomeo-ri, Mulguemeup, Yangsan, South Korea examined the anti-diabetic molecular mechanisms and anti-hyperglycemic effects on new anti-diabetic formula consisting of seven multi-herbs - hypoglycaemic candidates including *Panax ginseng* C. A. Meyer, *Pueraria lobata*, *Dioscorea batatas* Decaisne, *Rehmannia glutinosa*, *Amomum cadamomum* Linné, *Poncirus fructus* and *Evodia officinalis*, by using both 4 cell models such as human embryonic kidney (HEK) 293 (CRL- 1573), 3T3-L1 pre-adipocytes (CL-173), HepG2 hepatocytes (HB-8065) and C_2C_{12} skeletal myoblast cells (CRL-1772), and male C57BL/KsJ-db/db mice of genetic animal model for type 2 diabetes. First, on peroxisome proliferator-activated receptor γ (PPARγ) agonists, a compound that binds to receptors and activates intracellular signals - agonists of nuclear receptor PPARγ are therapeutically used to combat hyperglycaemia associated with metabolic syndrome and type 2 diabetes [25].

First, HEK293 cells were transfected with paraformaldehyde (pFA)-peroxisome proliferator-activated receptor gamma (PPARγ) and pFR-Luc (UAS-Gal4-luciferase), and then treated with extract (5 μg/mL), rosiglitazone (Rosi. **17**) (Figure 7) (10 μM), or macelignan (Mace. **18**) (Figure 7) (10 μM) for 24 hrs. Their relative luciferase activity (% of non treatment) was around 100 for non (control), around 1450 for diabetes treatment agent (insulin resistance improvement) rosiglitazone (Rosi. **17**) (10 µM), around 750 for macelignan (Mace. **18**) (10 µM) and around 650 for extract (5 µg/mL), respectively. Each extract increased the ligand-binding activity of PPARγ.

From these results it was evident that the extract increased PPARγ-dependently luciferase activity similar to that (luciferase activity) of a well-known PPARγ agonist rosiglitazone (Rosi. **17**), and PPARα/γ dual functional agonist used as positive control throughout the experiments with macelignan (Mace. **18**).

Second, the differentiated 3T3-L1 adipocytes (CL-173) were transfected with plasmid 3X PPREs-thymidine kinase-Luc (plasmid PPRE X3-TK-luc) and treated with extract (5 µg/mL), rosiglitazone (Rosi. **17**) (10 µM), or macelignan (Mace. **18**) (10 µM) for 24 hrs. Their relative luciferase activity (% of non treatment) was around 100 for non (control), around 1400 for rosiglitazone (Rosi. **17**) (10 µM), around 740 for macelignan (Mace. **18**) (10 µM) and around 640 for extract (5 µg/mL), respectively. Each extract induced transcriptional activity of PPARγ.

In order to further confirm PPARγ agonist potential of the extract, transient transfections were performed in differentiated 3T3-L1 adipocytes (CL-173) with thymidine kinase-luciferase vector (tk-luciferase vector) containing PPAR-responsive elements (PPREs) and then treated with the extract. The treatment of extract (5 µg/mL), rosiglitazone (Rosi. **17**), and macelignan (Mace. **18**) stimulated peroxisome proliferator response element (PPRE)-dependent luciferase activities in transfected cells.

Third, a fat dyeing method using azo dye for confirming differentiation of adipose precursor cells to adipocytes oil red O staining (lipid stain method) was measured after induction of differentiation of 3T3-L1 adipocytes in medium containing 0.1% DMSO (control), extract (5 µg/mL), rosiglitazone (Rosi. **17**) (1 µM), or macelignan (Mace. **18**) (10 µM) for 7 days. Each extract induced adipogenesis.

To provide biological evidence that the extract is a PPARγ ligand of substances specifically binding to receptor protein PPARγ. It was further investigated for adipocyte differentiation and expression of adipocyte marker gene in the extract-treated 3T3-L1 adipocytes. The extract-treatment led to significant increase in the formation of lipid droplets of fat-filled structure inside cells in similar to rosiglitazone (Rosi. **17**) and macelignan (Mace. **18**).

Fourth, the differentiated 3T3-L1 cells were treated with extract (5 μg/mL), rosiglitazone (Rosi. **17**) (10 μM), or macelignan (Mace. **18**) (10 μM) for 24 hrs. The expression of mRNAs was estimated using quantitative real-time method of amplifying complementary DNA (cDNA) prepared from reverse transcriptase (RT) from messenger RNA (mRNA) by polymerase chain reaction (PCR) reverse transcription-polymerase chain reaction (RT-PCR), and the results were expressed as mRNA levels relative to 0.1% DMSO (control). Their relative expression (% of non treatment) was around 100 for 0.1% DMSO (control), around 440 for rosiglitazone (Rosi. **17**) (10 μM), around 220 for macelignan (Mace. **18**) (10 μM) and around 210 for extract (5 μg/mL), respectively. Each extract increased PPARγ-target gene aP2 expression in differentiated 3T3-L1 adipocytes.

Additionally, the extract increased expression of adipose fatty acid-binding protein (aP2) suggesting that the extract was one of effective PPARγ agonists.

Second, on the effect on AMP-activated protein kinase (AMPK) activation in C_2C_{12} skeletal myoblast cells (CRL-1772), *first*, C_2C_{12} skeletal myoblast cells (CRL-1772) were treated with aminoimidazole-4-carboxamide-1-β-d-ribofuranoside (AICAR. **19**) (Figure 7) (1 mmol/L), extract (5 μg/mL), or macelignan (Mace. **18**) (10 μM)) for 24 hrs. The phosphorylated AMPK was examined by Western blot analysis. Their relative amount (% of non treatment) was around 100 for 0.1% DMSO (control), around 3300 for aminoimidazole-4-carboxamide-1-β-d-ribofuranoside (AICAR. **19**), around 450 for macelignan (Mace. **18**), and around 600 for extract, respectively. The extract increased AMPK phosphorylation.

Second, the expression of β-actin (**20**) was estimated using quantitative real-time RT-PCR as follows: Their relative expression (% of non treatment) of acyl-CoA synthetase (ACS) was around 100 for 0.1% DMSO (control), around 265 for aminoimidazole-4-carboxamide-1-β-d-ribofuranoside (AICAR. **19**) (1 mmol/L), around 170 for macelignan (Mace. **18**) (10 μM) and around 165 for extract (5 μg/mL), respectively. Their relative expression (% of non treatment) of carnitine palmitoyltransferase-1 (CPT-1) was around 100 for 0.1% DMSO (control), around 275 for aminoimidazole-4-carboxamide-1-β-d-ribofuranoside (AICAR. **19**) (1 mmol/L), around 150 for macelignan (Mace. **18**) (10 μM) and around 140 for extract (5 μg/mL), respectively. Thus, the extract increased mRNA expression of acyl-CoA synthetase (ACS) of enzyme catalyzing the reaction to acylcarnitine and coenzyme A (CoA) using long-chain acyl CoA and carnitine as substrates and carnitine palmitoyl-transferase-1 (CPT-1) of mitochondrial outer membrane enzyme catalyzing the binding of fatty acid and carnitine. This suggested that the extract activated AMP-activated protein kinase (AMPK).

Third, it is known that adipose tissue secretes inflammatory cytokines, and tumor necrosis factor-alpha (TNF-α) inhibits uptake of glucose into cells and decreases sensitivity to insulin [26]. Based on this fact, it was investigated whether the extract possessed anti-inflammatory effects, including the inhibitory effects of the extract on an activating enzyme of transcription factor nuclear factor-kappa B (NFκB) inhibitory kappa B kinase β (IKKβ)/a protein complex acting as a transcription factor nuclear factor-kappa B (NFκB) signal-transduction in human HepG2 cells (human hepatoblastoma G2 cells) treated with tumor necrosis factor-alpha (TNF-α) using nuclear factor-kappa B (NFκB) response element containing reporter NFκB-luciferase reporter because inflammatory processes play potential roles in the pathogenesis of insulin resistance.

First-their relative activity (% of non treatment) on the inhibition of a light emitting enzyme luciferase activity in TNF-α treated HepG2 cells was around 100 for 0.1% DMSO (control), around 1800 for 0.1% DMSO (control) with TNF-α (10 ng/mL), around 550 for antidiabetic drug for improvement of insulin resistance rosiglitazone (Rosi. **17**) (10 μM) with TNF-α (10 ng/mL), around 300 for macelignan (Mace. **18**) (10 μM) with

TNF-α (10 ng/mL), and around 250 for extract (5 μg/mL) with TNF-α (10 ng/mL), respectively. As a result, the extract prevented the increase of TNF-α-stimulated luciferase activity in TNF-α treated HepG2 cells. In summary, TNF-α treatment increased the nuclear factor-kappa B (NFκB)-dependent luciferase activity (p = 0.001 vs non-treatment), and the extract effectively prevented this increase (p = 0.034 vs TNF-α treatment).

Second-inhibitory kappa B kinase β (IKKβ) (IκBα) was measured by Western blot analysis. The extract increased I-kappa B protein (IkB) level. The extract increased the IkappaB-alpha (IkBa) level that was reduced by TNF- α treatment, which was consistent with rosiglitazone (Rosi. **17**) and macelignan (Mace. **18**). These results indicated that the extract exerted anti-inflammatory effects.

Fourth, it is known that endoplasmic recticulum (ER) stress plays central role in the development of insulin resistance and diabetes by impairing insulin signaling through c-Jun NH_2-terminal kinase (JNK) activation [27].

Based on the above fact, it was investigated that whether the extract was effective against attenuation of endoplasmic recticulum (ER) stress.

First-the inhibitory effect of endoplasmic recticulum (ER) stress by the extract was examined by using luciferase activity of enzyme-linked immunosorbent assay (ELISA) response element containing reporter in HepG2 cells treated with endoplasmic recticulum (ER) stress inducer - an endoplasmic reticulum Ca^{2+} pump inhibitor thapsigargin (**21**) (Figure 7).

The relative activity (% of non treatment) of the extract on inhibition of thapsigargin (**21**)-stimulated luciferase activity in thapsigargin (**21**)-treated HepG2 cells was around 100 for 0.1% DMSO (control), around 1450 for 0.1% DMSO (control) with thapsigargin (**21**) (10 ng/mL), around 1250 for antidiabetic drug for improvement of insulin resistance rosiglitazone (Rosi. **17**) (10 μM) with thapsigargin (**21**) (10 ng/mL), around 350 for macelignan (Mace. **18**) (10 μM) with thapsigargin (**21**) (10 ng/mL), and around 400 for extract (5 μg/mL) with thapsigargin (**21**) (10 ng/mL), respectively. As a result, while thapsigargin (**21**) treatment increased the enzyme-linked immunosorbent assay (ELISA)-dependent luciferase activity ($p = 0.001$ vs nontreatment), the extract effectively blocked thapsigargin (**21**)-mediated stimulation ($p = 0.039$ vs thapsigargin treatment).

Second-the relative activity (% of non treatment) of the extract using a ER stress indicator GRP78 on inhibition of thapsigargin (**21**)-stimulated luciferase activity in thapsigargin (**21**)-treated HepG2 cells was around 100 for 0.1% DMSO (control), around 530 for 0.1% DMSO (control) with thapsigargin (**21**) (10 ng/mL), around 460 for an antidiabetic drug for improvement of insulin resistance rosiglitazone (Rosi. **17**) (10 μM) with thapsigargin (**21**) (10 ng/mL), around 160 for macelignan (Mace. **35**) (10 μM) with thapsigargin (**21**) (10 ng/mL), and around 210 for extract (5 μg/mL) with thapsigargin (**21**) (10 ng/mL), respectively.

Third-the relative activity (% of non treatment) of the extract using ER stress indicator p-eIF2a on the inhibition of thapsigargin (**21**)-stimulated luciferase activity in thapsigargin (**21**)-treated HepG2 cells was around 100 for 0.1% DMSO (control), around 280 for 0.1% DMSO (control) with thapsigargin (**21**) (10 ng/mL), around 240 for an antidiabetic drug for improvement of insulin resistance rosiglitazone (Rosi. **17**) (10 μM) with thapsigargin (**21**) (10 ng/mL), around 70 for macelignan (Mace. **18**) (10 μM) with thapsigargin (**21**) (10 ng/mL), and around 80 for extract (5 μg/mL) with thapsigargin (**21**) (10 ng/mL), respectively.

As a result, when two ER stress indicators - GRP78 and p-eIF2a - were examined in thapsigargin (**21**)-treated HepG2 cells, the extract treatment inhibited significantly the increase of these two indicators by thapsigargin (**21**) ($p = 0.045$ vs thapsigargin (**21**) treatment). From these facts, it was demonstrated that the extract exerted protective effects against ER stress.

Fifth, in a study to examine the *in vivo* anti-diabetic effects of the extract on diabetes, rosiglitazone (Rosi. **17**) (10 mg/kg), macelignan (Mace. **18**) (15 mg/kg) or extract (150 mg/kg) was orally administrated to diabetes C57BL/KsJ-(db/db) mice every day for three weeks, and effects of the extract were compared to that (body weight) of rosiglitazone (Rosi. **17**) and macelignan (Mace. **18**).

First, on the effects on body weight change and fasting blood glucose (FBG) in diabetes C57BL/KsJ-(db/db) mice, the body weight (g) of before and after was around 39 (before) and around 38 (after) for control, around 39 (before) and around 47 (after) for rosiglitazone (Rosi. **17**) (10 mg/kg), around 39 (before) and around 39.5 (after) for macelignan (Mace. **18**) (15 mg/kg), around 39.5 (before) and around 39.5 (after) for extract (150 mg/kg), respectively. As a result, the extract-treatment showed almost same body weights in diabetes C57BL/KsJ-(db/db) mice, before and after treatment. On the other hand, rosiglitazone (Rosi. **17**)-treated mice showed significantly higher body weights when compared to those (body weights) of other two – macelignan (Mace. **18**) and extract ($p = 0.001$ vs control).

Second, on fasting blood glucose level in diabetes C57BL/KsJ-(db/db) mice after 3-week treatment, fasting blood glucose (FBG) level (mg/dL) of before and after was around 250 (before) and around 810 (after) for control, around 300 (before) and around 310 (after) for rosiglitazone (Rosi. **17**) (10 mg/kg), around 300 (before) and around 580 (after) for macelignan (Mace. **18**) (15 mg/kg), around 300 (before) and around 700 (after) for extract (150 mg/kg), respectively. As a result, fasting blood glucose (FBG) levels at the baseline (day 0) did not differ among 4 groups – control, rosiglitazone (Rosi. **17**), macelignan (Mace. **18**), and extract. However, levels of the extract-treated group after 3 weeks were significantly lower when compared to that (fasting blood glucose (FBG) level) of diabetes C57BL/KsJ-(db/db) mice control group ($p = 0.022$ vs control). The other two groups (fasting blood glucose levels (FBG)) treated with rosiglitazone (Rosi. **17**) ($p = 0.001$ vs control) and macelignan (Mace. **18**) ($p = 0.002$ vs control) were significantly lower when compared to that (fasting blood glucose (FBG) level) of diabetes C57BL/KsJ-(db/db) mice control group. Blood glucose levels of the extract-treated diabetes C57BL/KsJ-(db/db) mice showed significant reduction by about 15% when compared to that (blood glucose level) of the control.

Sixth, regarding the effects on postprandial blood glucose (PPG) level and insulin sensitivity in diabetes C57BL/KsJ-(db/db) mice, to assess the glucose homeostasis and insulin sensitivity in diabetes C57BL/KsJ-(db/db) mice treated with the extract, glucose tolerance and insulin tolerance tests at 120 mins before end of the experiment were measured.

First, the extract within 120 mins showed significant reduction of postprandial blood glucose (PPG) level ($p = 0.001$ *vs* control) similar to two positive controls - rosiglitazone (Rosi. **17**) ($p = 0.003$ *vs* control) and macelignan (Mace. **18**) ($p = 0.004$ *vs* control) - when compared with that (blood glucose level) of diabetes C57BL/KsJ-(db/db) mice control groups.

Second, the insulin tolerance tests also showed that the reduction of postprandial blood glucose (PPG) level in response to insulin (**22**) was much greater in the extract-treated mice when compared to that (reduction of postprandial blood glucose (PPG) level) in untreated diabetes C57BL/KsJ-(db/db) mice ($p = 0.002$ *vs* control).

Thus, it was clear that the extract treatment affected not only regulation of postprandial blood glucose (PPG) level, but also enhanced insulin sensitivity.

Seventh, on effects of the extract against plasma lipids including free fatty acids (FFAs) level and plasma triglycerides (plasma TG) level, and total cholesterol (TC) level in diabetes C57BL/KsJ-(db/db) mice, 3 plasma lipid levels were around 2.28 (mmol/L) (FFAs), around 296.2 (mg/dL) (plasma TG), around 146.1 (mg/dL), 146.1 (mg/dL) (total cholesterol. TC) for control, around 0.94 (mmol/L) (FFAs), around 109.4 (mg/dL) (plasma TG), around 181.9 (mg/dL) (TC) for rosiglitazone (Rosi. **17**) (10 mg/kg), around 1.70 (mmol/L) (FFAs), around 259.0 (mg/dL) (plasma TG), around 110.0 (mg/dL) (TC) for macelignan (Mace. **18**) (15 mg/kg), and around 1.75 (mmol/L) (FFAs), around 217.9 (mg/dL) (plasma TG), and around 119.4 (mg/dL) (TC) for extract (150 mg/kg), respectively.

The extract-treatment significantly decreased plasma free fatty acids (FFAs) ($p = 0.021$ *vs* control), plasma triglycerides (plasma TG) ($p = 0.012$ *vs* control) and total cholesterol (TC) ($p = 0.003$ *vs* control) levels of diabetes C57BL/KsJ-(db/db) mice control when compared to those (FFAs, plasma TG, TC) of untreated diabetes C57BL/KsJ-(db/db) mice, at the end of experiment.. These results demonstrated that this decrease in plasma lipids might contribute to the improvement of severe diabetes at least partially, as lipolysis and circulating free fatty acids increased under insulin (**39**) resistance conditions.

Eighth, generally it is known that blood glycosylated hemoglobin A_{1c} (HbA$_{1c}$. HbA1c) is hemoglobin (Hb) in which glucose is nonenzymatically bound, and HbA1c is used as a criterion for judging presence or absence of diabetes or as indicator of the state of glycemic control since it represents average blood glucose level in the past one or two months. The higher glucose concentration (= blood glucose level) in blood, longer exposure time of hemoglobin (Hb) to glucose results in the acceleration of HbA1c progresses. Therefore, HbA1c shows a positive correlation with blood sugar level.

Glucagon is hormone with hyperglycemic action in pancreas. Insulin is made of beta cells of pancreas, in addition to a substance called C-peptide immunoreactivity (CPR. C-peptide). Proinsulin (precursor of insulin) synthesized in β cells, a substance underlying insulin, is broken down by enzymes to generate one molecule of each of insulin and C-peptide. Therefore, the same amount of insulin and C-peptide is secreted.

In type 2 diabetes, a decrease in insulin secretion results in low C-peptide immunoreactivity (CPR). The molar ratio of insulin/glucagon (I/G) relates to the biologic opposition of insulin and glucagon exertion upon liver and adipose tissue suggesting that the relative concentrations of these two hormones which perfuse these tissues may determine their net nutrient balance [28].

Regarding effects of the extract on concentrations of blood and plasma biomarkers in diabetes C57BL/KsJ-(db/db) mice, the concentrations of blood and plasma biomarkers were around 10.7 (%) (HbA1c), around 1.48 (ng/mL) (plasma insulin), around 0.37 (ng/mL) (glucagon), around 3.12 (ng/mL) (CPR), around 4.68 (I/G molar ratio) for control, around 7.40 (%) (HbA1c), around 3.43 (ng/mL) (insulin), around 0.32 (ng/mL) (glucagon), around 4.76 (ng/mL) (CPR), around 9.67 (I/G molar ratio) for rosiglitazone (Rosi. **17**) (10 mg/kg), around 10.8 (%) (HbA1c), around 1.52 (ng/mL) (insulin), around 0.23 (ng/mL) (glucagon), around 4.14 (ng/mL) (CPR), around 6.74 (I/G molar ratio) for macelignan (Mace. **18**) (15 mg/kg), around 9.3 (%) (HbA1c), around 3.15 (ng/mL) (insulin), around 0.21 (ng/mL) (glucagon), around 4.79 (ng/mL) (CPR), and around 14.2 (I/G molar ratio) for extract (150 mg/kg), respectively.

As a result, the extract-treated diabetes C57BL/KsJ-(db/db) mice showed significantly lower blood glycosylated hemoglobin (HbA1c) level when compared to that ((HbA1c) of diabetes C57BL/KsJ-(db/db) mice control (p = 0.002 *vs* control). Both plasma insulin (p = 0.042 *vs* control) and CPR levels (p = 0.038 *vs* control) were significantly higher in the extract-treated diabetes C57BL/KsJ-(db/db) mice when compared to those (plasma insulin level and CPR level) in diabetes C57BL/KsJ-(db/db) mice control. However, glucagon level in the extract-treated diabetes C57BL/KsJ-(db/db) mice were significantly lower when compared to that (glucagon level) of diabetes C57BL/KsJ-(db/db) mice control (p = 0.018 *vs* control). The extract-treatment significantly improved molar ratio of insulin/glucagon (I/G) when compared to that (I/G molar ratio) of diabetes C57BL/KsJ-(db/db) mice control.

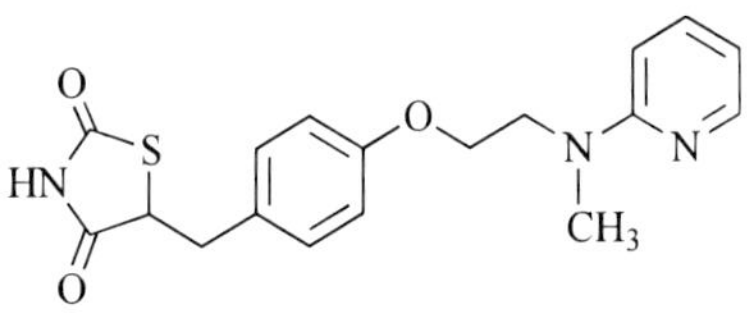

rosiglitazone (Rosi. **17**)

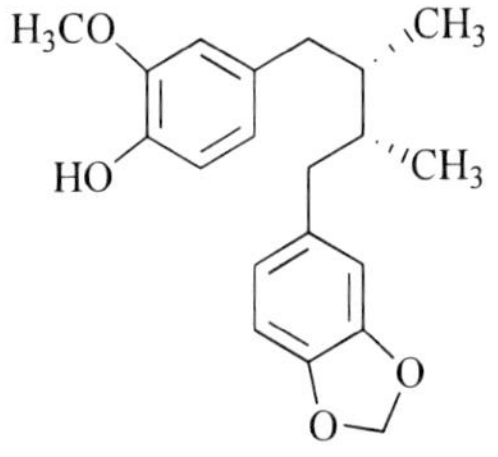

macelignan (Mace. **18**)

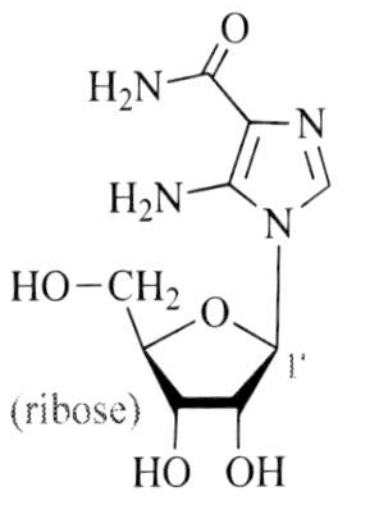

aminoimidazole-4-carboxamide-1-β-D-ribofuranoside (AICAR. **19**)

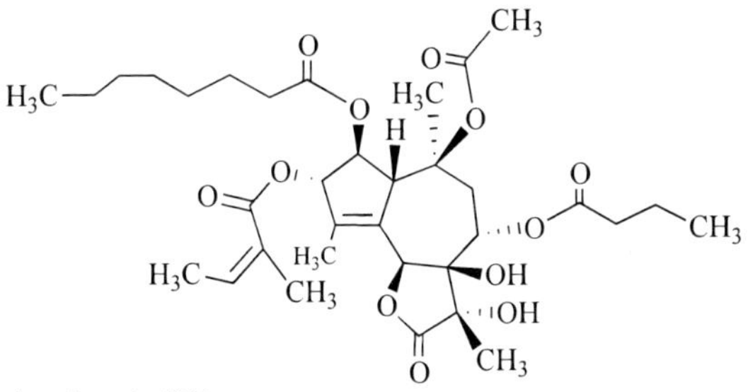

thapsigargin (**21**)

Figure 7. Rosiglitazone (Rosi. **17**), macelignan (Mace. **18**), aminoimidazole-4-carboxamide-1-β-D-ribofuranoside (AICAR. **19**), and thapsigargin (**21**).

From the above results it was evident that aqueous extract of these 7 hypoglycemic herbs including *Panax ginseng* C. A. Meyer was very effective in the improvement and prevention of type 2 diabetes. These seven herbal mixed extract was thought to exert high antidiabetic effect by synergistic or additive action among phytochemicals contained in each seven herbs [29].

2.1.7. Antiviral Effects

In 1996, a group from Department of Pharmacology, University of Milan, Italy examined the antiviral effects of standardised G115 ginseng extract (100 mg) of a standardized *Panax ginseng* root extract with a higher immune response in vaccination against influenza, using total of 227 volunteers for a period of 12 weeks.

As a result, first, the frequency of influenza or common cold between weeks 4 and 12 was 42 cases in placebo group. However, only 15 cases in standardised G115 ginseng extract group showed higher immune response when compared to that (the frequency) of placebo group.

Second, antibody titres by week 8 rose to an average of 171 units in placebo group and an average of 272 units in standardised G115 ginseng extract group ($p < 0.0001$). Third, natural killer (NK) activity levels at weeks 8 and 12 were nearly twice as high in standardised G115 ginseng extract group when compared to that (NK activity levels) of placebo group ($p < 0.0001$). Fourth, in all 227 volunteers, values of 24 safety parameters showed no significant differences between the end and beginning of 12-week study in either of these groups. There were only 9 adverse events in this study, and among these, insomnia was the principal adverse effect [30].

2.1.8. Treatment of Erectile Dysfunction (E.D.) or Impotence

On the treatment of erectile dysfunction (E.D.), in 1995, a group from Severance Institute of Andrology Research, Yonsei University, College of Medicine, Seoul, Korea examined both the efficacy and natural drug development without complications against erectile dysfunction (E.D.) by the comparison of standardized *Panax ginseng* (Korean red ginseng) root extract (600 mg orally three times daily) and placebo with serotonin (5-

hydroxytryptamine. 5-HT. **23**) (Figure 8) antagonist and serotonin antagonist reuptake inhibitor (SARI) trazodone (**24**) (Figure 8) used as a therapeutic agent for erectile dysfunction (E.D.), using total of 90 patients with 30 patients in three each group - placebo group (control), standardized *Panax ginseng* (Korean red ginseng) root extract group, and trazodone (**24**) group.

First, the erectile dysfunction (E.D.) symptomatic change - frequency of intercourse, premature ejaculation, and morning erections after treatment - were not changed in all three groups.

First-on changes in early detumescence and erectile parameters, the changes in early detumescence and erectile parameters (e.g., penile rigidity and girth, libido and patient satisfactions) in standardized *Panax ginseng* (Korean red ginseng) root extract group were significantly higher when compared to those (the changes) of other two groups – placebo group (control) and trazodone (**24**) group ($p < 0.05$). Concretely, the overall therapeutic efficaciy on erectile dysfunction (E.D.) was 60% for standardized *Panax ginseng* (Korean red ginseng) root extract group, and 30% for placebo group (control) and trazodone (**24**) group, respectively, showing more beneficial effects of *Panax ginseng* root extract from statistical observation ($p < 0.05$).

Second, erectile dysfunction (E.D.) complete remission was not reported, while partial responses were reported. However, the aggravation of erectile dysfunction (E.D.) symptoms also was not reported.

Additionally, audiovisual stimulation penogram (AVS-penogram) - a recording of penile hemodynamic changes during natural erection after audiovisual erotic stimulation as initial screening test for impotent patients - did not change after the oral administration of standardized *Panax ginseng* (Korean red ginseng) root extract. However, when impotent patients were orally administered standardized *Panax ginseng* root extract for long time period, the cumulative effect on penile blood flow might be observed.

From these results (First - Second) above, the oral administration of standardized *Panax ginseng* (Korean red ginseng) root extract to impotent patients showed significantly higher improvement effect when compared to that (improvement effect) of placebo group (control) or trazodone (**24**) group. The relieving effect of standardized *Panax ginseng* (Korean red ginseng) root extract (600 mg orally three times daily) for impotent patients will be due to saponins extracted from *Panax ginseng*, because the mechanism of action on penile corpus cavernous smooth muscle of erectile tissues was confirmed by organ chamber or nitric oxide titration for saponins. Additionally, it was shown that possible finding in treatment of erectile dysfunction (E.D.) arises from active saponins extracted from Korean red ginseng, bringing hopes to erectile dysfunction (E.D.) patients [31].

In 2007, a group from Sector of Sexual Medicine, Division of Urological Clinic of Sao Paulo University, Sao Paulo, Brazil examined the treatment efficacy of Korean red ginseng (KRG) extracts (either 1000 mg 3 times daily of KRG or a placebo) for erectile dysfunction (E.D.) patients using a total of 60 mild or mild to moderate erectile dysfunction (E.D.) patients, through five-item version of International Index of Erectile Function (IIEF-5). The erectile dysfunction (E.D.) patients were divided into two groups with 30 patients in each group and randomized into a 12-week double-blind protocol, and received either 1000 mg Korean red ginseng (KRG) extracts or placebo (capsule containing starch with KRG flavor) 3 times daily.

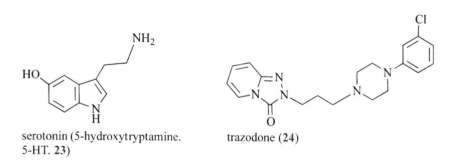

Figure 8. Serotonin (5-hydroxytryptamine. 5-HT. **23**) and trazodone (**24**).

First, *first*-average baseline score of IIEF-5 in Korean red ginseng (KRG) extract group (1000 mg 3 times daily) was around 16.4. The IIEF-5 score increased to around 21.0 after treatment ($p < 0.01$). The average baseline IIEF-5 score in placebo group was around 17.0 and changed to around 17.7 after treatment, respectively ($p > 0.05$). *Second*, 20 patients (66.6%) in Korean red ginseng (KRG) group showed improved erectile dysfunction (E.D.) for answers to global efficacy question (Q1, Q2, Q3, Q4, Q5, and Q15 (GAQ) ($p < 0.01$). However there was no reported improvement of erectile dysfunction (E.D.) in placebo group. Korean red ginseng (KRG) group showed more high rigidity, maintenance of erection, and penetration when compared to those (rigidity, maintenance of erection, and penetration) in placebo group ($p < 0.01$).

The IIEF-5 score after KRG extract treatment in Korean red ginseng (KRG) group significantly improved in total IIEF-5 score, showing evidence for questions (What are the questions?) 3 ($p < 0.001$) and 5 ($p < 0.0001$), respectively) when compared to that (IIEF-5 score) of placebo group.

Second, on the changes of average serum testosterone, prolactine, cholesterol, high-density lipoprotein (HDL), and low-density lipoprotein (LDL) after KRG extract treatment, there was no difference between Korean red ginseng (KRG) group and placebo group.

No patients presented with hypogonadism.

3 Erectile dysfunction (E.D.) patients in Korean red ginseng (KRG) group showed minor side effects - headache and insomnia.

From these results (First and Second) above, Korean red ginseng (KRG) group was effective and safe for treatment for erectile dysfunction (E.D.). Here, it is known that *Panax ginseng* rectified the erectile dysfunction (E.D.) [32]. *Panax ginseng* enhanced the vasodilator action and platelet aggregation inhibitory action by nitric oxide arisen from nitric oxide (NO) synthase in endothelium, and worked as antioxidant and ensured a protective role [33, 34]. Therefore, it was suggested that the enhanced NO synthase in corpus cavernous could improve erection and therapeutic action on erectile dysfunction (E.D.) [35].

2.1.9. Enhancement of Physical and Psychological Performance Capacity (Improvement of Lack- Stamina)

2.1.9.1. Actoprotectors and Adaptogens

Actoprotectors are compounds that enhance body stability against physical loads without increasing oxygen consumption or heat production, and aid to improve human's physical and mental efficiency [36]. Generally, it is thought that actoprotectors enhance effects of nootropic drugs including antiepileptic drug with antimyoclonic effect of piracetam which is brain function regulating drug, strengthening cognitive function and preventing brain aging, psychostimulant, and adaptogens. Adaptogens are natural herbs (e.g., *Hypericum perforatum*. St. John's wort) that work to enhance resistance to stress such as trauma, anxiety, physical fatigue and so on.

Actoprotectors are mainly classified into three (Figure 9) as follows:

1. **Benzthioimidazoles:** bemitil (2-ethylthiobenzimidazole hydrobromide. **25**), ethomersol (**26**).
2. **Adamantanes** (crosslinked hydrocarbon having a diamond-like structure): bromantane (*N*-(2-adamantil)-*N*-(*p*-bromophenyl)-amine. **27**), chlodantan (**28**), ademol (**29**).
3. **Miscellaneous compounds:** ginseng ginsenosides such as ginsenoside Rb1 (**6**), thiazoloindoles, 3-hydroxypyridines, nicotinic acids, 1-oxa-4aza-2-silacyclanes, chitosans (glucosamine polysaccharides), and others [37].

2.1.9.2. Effects of Ginseng Saponins (Ginsenosides) on Animal Adrenal Chromaffin Cells

In 1999, a group from Department of Pharmacology, School of Medicine, Iwate Medical University, Morioka, Japan examined the effects of 4 representative ginseng saponins (ginsenosides) of three groups – *panaxadiol group* (e.g., ginsenoside-Rb1 (**6**), ginsenoside-Rg3 (**30**); *panaxatriol group* (e.g., ginsenoside-Rg2 (**31**) (Figure 10a); *oleanolic acid group* (e.g., ginsenoside-Ro (**32**)) (Figure 10b) on the secretion of catecholamines from bovine and guinea-pig adrenal chromaffin cells and on the contraction of ileum in guinea-pig induced by various receptor stimulants - γ-amino- butyric acid (GABA. **33**), histamine (His. **34**), bradykinin (BK. **35**), angiotensin II, neurotensin (NT) and muscarine (**36**) (Figure 11).

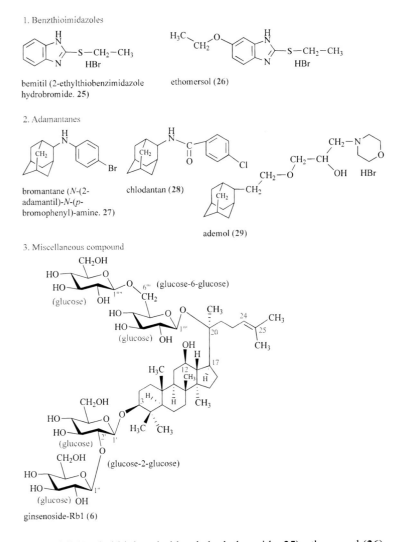

Figure 9. Bemitil (2-ethylthiobenzimidazole hydrobromide. **25**), ethomersol (**26**), bromantane (*N*-(2-adamantil)-*N*-(*p*-bromophenyl)-amine. **27**), chlodantan (**28**), ademol (**29**), and ginsenoside Rb1 (**6**).

1. panaxadiol group: ginsenoside-Rb1 (**6**) and ginsenoside-Rg3 (**30**)

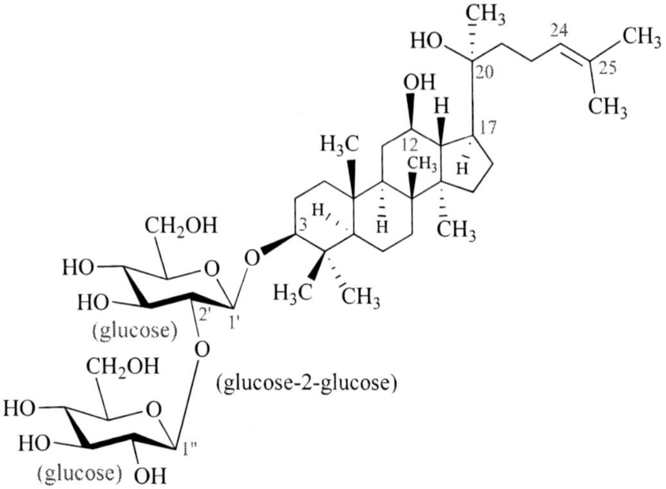

ginsenoside-Rg3 (**30**)

2. panaxatriol group: ginsenoside-Rg2 (**31**)

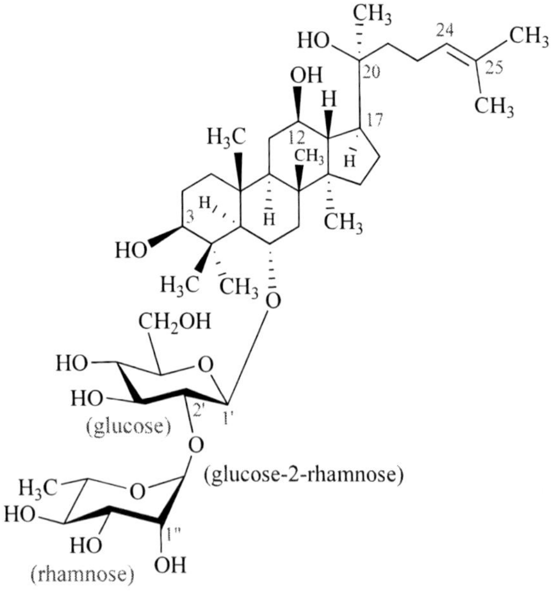

ginsenoside-Rg2 (**31**)

Figure 10a. *Panaxadiol Group:* ginsenoside-Rb1 (**6**) and ginsenoside-Rg3 (**30**); *panaxatriol Group:* ginsenoside-Rg2 (**31**).

3. *oleanolic acid group*: ginsenoside-Ro (**32**)

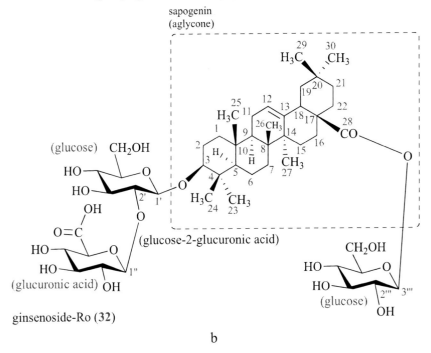

Figure 10b. *Oleanolic acid Group:* ginsenoside-Ro (**32**).

Chromaffin cells are a type of cells produced by nerve cells. Chromaffin cells generate neurohormones (neurotransmitters) that are used to send signals (neurotransmitters) to other cells thereby releasing them into the blood. Chromaffin cells make epinephrine (adrenaline) and norepinephrine (noradrenaline) and are found in two adrenal glands or in groups of nerve cells called ganglia.

First, regarding 50% inhibitory concentration (IC50) (µM) of 4 ginsenosides (ginsenoside-Rb1 (**6**), ginsenoside-Rg3 (**30**), ginsenoside-Rg2 (**31**), and ginsenoside-Ro (**32**)) on bovine adrenal chromaffin cells from 6 stimulants [acetylcholine (Ach. **36**) (Figure 11), γ-amino-butyric acid (GABA. **33**), histamine (His. **34**), angiotensin II (AT II), bradykinin (BK. **35**) and neurotensin (NT)-induced secretion of catecholamines, the IC50 (µM) was 4 (ginsenoside-Rg2 (**31**)) and 10 (ginsenoside-Rg3 (**30**)) for a neurotransmitter acetylcholine (Ach. **36**) (100 µM), 10 (ginsenoside-Rg2

(**31**)) and 26 (ginsenoside-Rg3 (**30**)) for γ-amino- butyric acid (GABA. **33**) (40 μM), - (no inhibition) (ginsenoside-Rg2 (**31**)) and 62 (ginsenoside-Rg3 (**30**)) for histamine (His. **34**) (10 μM), - (ginsenoside-Rg2 (**31**)) and 64 (ginsenoside-Rg3 (**30**)) for angiotensin II (AT II) (100 nM), - (ginsenoside-Rg2 (**31**)) and >100 (ginsenoside-Rg3 (**30**)) for bradykinin (BK. **35**) (10 nM), and - (ginsenoside-Rg2 (**31**)) and 28 (ginsenoside-Rg3 (**30**)) for neurotensin (NT) (20 μM), respectively (original Table 1). However, ginsenoside-Rb1 (**6**) and ginsenoside-Ro (**32**) had no more significant inhibition of catecholamine secretion when compared to that (inhibition of catecholamine secretion) of ginsenoside-Rg2 (**31**) and ginsenoside-Rg3 (**30**).

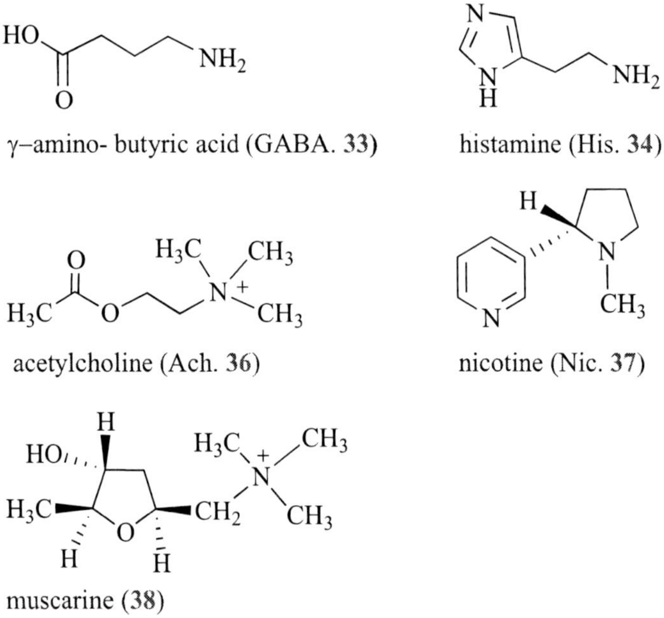

γ–amino- butyric acid (GABA. **33**)

histamine (His. **34**)

acetylcholine (Ach. **36**)

nicotine (Nic. **37**)

muscarine (**38**)

Figure 11. γ-Amino- butyric acid (GABA. **33**), histamine (His. **34**), acetylcholine (Ach. **36**), nicotine (Nic. **37**), and muscarine (**38**).

From these above results, among 6 stimulants - acetylcholine (Ach. **36**), γ-amino-butyric acid (GABA. **33**), histamine (His. **34**), angiotensin II (AT II), bradykinin (BK. **35**) and neurotensin (NT), the only two IC50s of ginsenoside-Rg2 (**31**) against acetylcholine (Ach. **36**) and γ-amino-butyric acid (GABA. **33**) were higher when compared to those (the two IC50s) of ginsenoside-Rg3 (**30**). Additionally, the IC50s of ginsenoside-Rg3 (**19**) against histamine (His. **34**), angiotensin II (AT II), bradykinin (BK. **35**) and neurotensin (NT) were more significant when compared to those (IC50s: no inhibitions) of ginsenoside-Rg2 (**31**).

Second, for the actions of ginsenosides with nicotinic acetylcholine receptors (nAChR), regarding the effects of ginsenoside-Rb1 (**6**), ginsenoside-Rg2 (**31**) and ginsenoside-Rg3 (**30**) on catecholamine secretion from guinea-pig chromaffin cells stimulated by nicotine (**37**) (Figure 11), percentage (%) of catecholamine secretion was 100 for nicotine (Nic. **37**) (100 mM) without ginsenoside (GS) (100 μM); around 6 for ginsenoside-Rg2 (**31**) (100 μM) + nicotine (Nic. **37**) (100 μM); around 85 for ginsenoside-Rb1 (**6**) (100 μM) + nicotine (Nic. **37**) (100 μM) and around 95 for ginsenoside-Rg3 (**30**) (100 μM) + nicotine (Nic. **37**) (100 μM), respectively.

From the above results it was evident that ginsenoside-Rg2 (**31**) (100 μM) significantly reduced nicotine (**37**)-induced secretion from chromaffin cells, whereas ginsenoside-Rb1 (**6**) and ginsenoside-Rg3 (**30**) did not affect the secretion.

Third, on the effects of ginsenosides (GS: 100 μM) on muscarine (**38**) (Figure 11) (3 nM–30 μM: final concentrations)-induced contraction of ileum in guinea-pig, the isolated ileum was preincubated without or with ginsenosides (GS: 100 μM). As a result, *first*-guinea-pig ileum contracts *via* stimulation of a variety of receptors. Muscarine (**38**) produced the contraction, dose-dependently (30 nM-3 μM).

Second-two ginsenoids (ginsenoside-Ro (**32**) and ginsenoside-Rg2 (**31**)) even at higher concentration (100 μM) scarcely affected muscarine (**38**)-induced contraction. On the other hand, ginsenoside-Rg3 (**30**) at concentrations (3-100 μM) gave parallel shift towards right of the concentration-response curve to muscarine (**38**).

Third-ginsenoside-Rb1 (**6**) at a higher concentration (100 μM) suppressed the contraction induced by the lower concentrations of muscarine (**38**) (30–300 nM).

Fourth, on the effects of ginsenosides (GS) on histamine (His. **34**)-induced contraction of ileum, histamine (His. **34**) also contracted ileum, dose-dependently (300 nM-10 μM).

From the results (First - Fourth) above, First, the effect of *oleanolic acid group* ginsenoside-Ro (**32**) did not show any inhibitory effect against *both* 6 stimulus (nicotine (Nic. **37**) (-: no effect), histamine (His. **34**) (-), angiotensin II (-), bradykinin (BK. **35**) (-), neurotensin (NT) (-) and γ-amino- butyric acid (GABA. **33**) (-) on bovine chromaffin cells, *and* against nicotine (Nic. **37**) (ND: not determined) on guinea-pig chromaffin cells, and against 2 stimulus muscarine (**38**) (-) and histamine (His. **34**) (-) on guinea-pig ileum.

Second, the effect of ginsenoside-Rb1 (**6**) of *panaxadiol group* showed slight inhibition (±) against nicotine (Nic. **37**) on bovine chromaffin cells and against muscarine (**38**) on guinea-pig ileum, and was not determined against nicotine (Nic. **37**) on guinea-pig chromaffin cells. Ginsenoside-Rb1 (**6**) did not show any inhibition (-) against other 5 stimulus (histamine (His. **34**), angiotensin II (AT II), bradykinin (BK. **35**), neurotensin (NT), γ-amino- butyric acid (GABA. **33**)) on bovine chromaffin cells and 2 stimulus (muscarine (**38**), histamine (His. **34**)) on guinea-pig ileum.

However, ginsenoside-Rg3 (**30**) of *panaxadiol group* showed strong inhibition (++) against 3 stimulus nicotine (Nic. **37**), neurotensin (NT) and γ-amino- butyric acid (GABA. **33**)) on bovine chromaffin cells, and strong inhibition (++) against 2 stimulus muscarine (**38**) and histamine (His. **34**) on guinea-pig ileum, respectively. Moreover, ginsenoside-Rg3 (**30**) showed inhibition (+) against 2 stimulus histamine (His. **34**) and angiotensin II (AT II) on bovine chromaffin cells, respectively.

Ginsenoside-Rg3 (**30**) showed slight inhibition (±) against bradykinin (BK. **35**) on bovine chromaffin cells, and showed no effect (-) against nicotine (Nic. **37**) on guinea-pig ileum.

Third, the effect of ginsenoside-Rg2 (**31**) of *panaxatriol group* showed strong inhibition (++) against nicotine (Nic. **37**) and γ-amino- butyric acid

(GABA. **33**) on bovine chromaffin cells, and against nicotine (Nic. **37**) on guinea-pig ileum, respectively. However, ginsenoside-Rg2 (**31**) did not show any inhibition (-) against other 4 stimulus (histamine (His. **34**), angiotensin II (AT II), bradykinin (BK. **35**), neurotensin (NT)) on bovine chromaffin cells, and 2 stimulus (muscarine (**38**), histamine (His. **34**)) on guinea-pig ileum.

Therefore, it was suggested that ginsenoside-Rg2 (**31**) of *panaxatriol group* in *Panax ginseng* was especially a high selective blocker of two ionotropic receptors - nicotinic acetylcholine receptor (nAChR) and γ-amino- butyric acid (GABA. **33**) receptor. However, ginsenoside-Rg3 (**30**) of *panaxadiol group* was not only a blocker of ionotropic receptors but also an antagonist of muscarinic receptor or histamine receptor [33].

2.1.9.3. Effects on Exercise Performance of Ginseng Ginsenosides (Steroid-Like Phytochemicals)

2.1.9.3.1. Effect on Exercise Performance of Ginseng Ginsenosides (Steroid-Like Phytochemicals) in Dry Roots of North American Ginseng (Panax Quinquefolius L) in Animal Studies

Since ancient times, mainly ginseng roots of two species of the genus Panax - *Panax ginseng* (Korean ginseng) and *Panax quinquefoliu* (American ginseng) of the family Araliaceae ginseng (*Panax* sp.) have been used as energy boosters and general tonics in both North America, China and surrounding countries. Ginseng root extracts increased general resistance against general hypoxia and cardiac ischemia of various noxious and stressful stimuli. Furthermore, the extract exerted stimulating effect for substrate metabolism by their alteration of lipid and carbohydrate mobilization as well as utilization [37, 38]. It was presumed that, both substrate availability and substrate utilization were crucial factors in restricting two potentialities of exercise performance and endurance. Therefore, this was a reason that ginseng might be useful as ergogenic aid in full marathon and exercise like sustained running. Actually, ginseng in animal studies significantly increased their endurance of swimming [39] or running on the rotating rod [40].

On the other hand, athletes have been using ginseng as an ergogenic aid for a long time. However there was no evidence of ginseng that supports the ergogenic aid for athletes by now because the effectiveness of ginseng on work performance might vary depending on the species of ginseng used in their individual studies, in addition to the differences in treatment duration and workload intensity [41].

It was most important to identify the bioactive components in ginseng which could enhance exercise performance. Generally, it was thought that ginseng root saponin (GS) fraction accounted for the bioactivity of ginseng root. A direct demonstration of ginseng root saponin (GS), a bioactive component which could enhance exercise performance was inconclusive. Additionally, since ginseng root saponin (GS) contained around 30 or more individual ginsenosides of ginseng saponin [42], it was very meaningful to isolate and identify the specific ginsenosides with exercise enhancing effect.

Based on the facts above, in 1998, a group from Department of Biological Sciences, University of Alberta, Edmonton, Canada examined the effectiveness of ginseng root saponin (GS) in dry roots of North American ginseng (*Panax quinquefolius* L.) on exercise performance and working mechanism as well as separation and identification of specific ginsenosides with exercise performance.

First, two main specific ginseng root saponins (GS) - protopanaxadiol type (group) saponin ginsenoside Rb1 (**6**) (4.2-5.1% *vs* dry root weight) and protopanaxatriol type (group) saponin ginsenoside Rg1 (**39**) (Figure 12) (0.48-0.52% *vs* dry root weight) were isolated from dry roots of North American ginseng (*Panax quinquefolius* L). Ginsenoside Rb1 (**6**) was of more than 95% purity when compared to that (purity) of ginsenoside Rg1 (**39**).

Second, on the effects of specific ginseng root saponin fraction (specific ginsenoside (GS) fraction) on exercise performance, *first*-the non-trained rats were intraperitoneally (*i.p.*) injected *acutely* the specific ginseng root saponin fraction (specific ginsenoside (GS) fraction) at 30 minute prior to exercise. As a result, the running times (min) after acute intraperitoneal injection (*i.p.*) by either sterile saline (SAL) and - 4 different doses (5, 10, 20, 80 mg/kg) of specific ginseng root saponin fractions (ginsenoside (GS) fraction) in non-trained adult male Sprague-Dawley (SD) rats was around 27.5 for sterile saline (SAL), around 25 for GS fraction (5 mg/kg), around 35 for GS fraction (10 mg/kg), around 32.5 for GS fraction (20 mg/kg) and around 28 for GS fraction (80 mg/kg), respectively. From the results, acute intraperitoneal injection *(i.p.)* of specific ginseng root saponin fraction (specific ginsenoside (GS) fraction) (5-80 mg/kg) did not cause any significant change in overall exercise performance of the non-trained rats.

Second, on the other hand, the non-trained rats were intraperitoneally (*i.p.*) injected 3 different doses (5, 10, 20 mg/kg/day) of specific ginseng root saponin fractions (ginsenoside (GS) fraction) every morning 10:00 hr for *4 days*. As a result, the running times (min) after *4 consecutive (chronic) days* of acute intraperitoneal injection *(i.p.)* by either sterile saline (SAL) and 3 different doses (5, 10, 20 mg/kg/day) of specific ginseng root saponin fraction (ginsenoside (GS) fraction) in non-trained adult male Sprague-Dawley (SD) rats was around 24.0 for sterile saline (SAL), around 26 for GS fraction (5 mg/kg/day), around 38 for GS fraction (10 mg/kg/day) and around 42.5 for GS fraction (20 mg/kg/day), respectively. These results demonstrated that the running times (min) increased dose-dependently with the dose of corresponding GS fraction, respectively.

Third, there was no change in maximal oxygen uptake (VO_{2max}) after specific ginsenoside (GS) fraction administration. However, the specific ginsenoside (GS) fraction-administrated rats tended to stay for a longer period of time in submaximal oxygen uptake ($VO_{2submax}$) level when compared to that ($VO_{2submax}$) of control sterile saline (SAL) rats.

Third, on the effects of specific ginsenoside (GS) fraction for substrate metabolism during exercise, to evaluate the physiological effects of specific ginsenoside (GS) fraction administration on exercise performance and their changes for 4 consecutive days of ginsenoside (GS) fraction administration on plasma substrate profile in three stages – before exercise (resting), during workload and after exhaustive exercise were examined.

The effects of 4 consecutive (chronic) days of acute intraperitoneal injection (*i.p.*) with either sterile saline (SAL) (1 mL/kg/day) or specific ginsenoside (GS) fraction (20 mg/kg/day) on time to reach 30% and 70% maximal oxygen uptakes (VO_{2max}), and exhaustion and the associated changes in plasma glucose, free fatty acids (FFAs), and lactate (**40**) (Figure 13) concentrations in rats at the corresponding time points were as follows:

First, the plasma glucose (**41**) (Figure 13) level (mg/dL) was around 108.2 (SAL) and around 114.2 (GS) for the resting level (prior to workload), around 104.5 (SAL) and around 113.9 (GS) for 30% maximal oxygen uptakes (30% VO_{2max}), around 93.3 (SAL) and around 111.5 (GS) for 70% maximal oxygen uptakes (70% VO_{2max}) and around 71.5 (SAL) and around 101.4 (GS) for exhaustion, respectively. From the above results, plasma glucose (**41**) level (mg/dL) of sterile saline-treated group decreased most steadily with increasing exercise workload. On the other hand, plasma glucose (**41**) level (around 101.4 mg/dL) of specific ginsenoside (GS) fraction-treated group after prolonged exercise was only slightly lower when compared to that (plasma glucose (**41**) level (around 114.2 mg/dL) of the resting level throughout the exercise period.

Second, plasma free fatty acids (FFA) levels (around 880.3 microequivalents per liter (µEq/L)) of specific ginsenoside (GS) fraction-treated group at rest (prior to workload), excluding glucose (**41**) and lactate (**40**), were significantly higher when compared to that (FFA level (around 516.6)) of sterile saline (SAL) control group at rest (prior to workload). Additionally, even though the FFA levels (µEq/L) of sterile saline (SAL)-treated group significantly decreased at around 516.6 (at rest), around 207.2 (30% maximal oxygen uptake (30% VO_{2max})), around 182.0 (70% maximal oxygen uptake (70% VO_{2max})), around 127.9 (exhaustion) during exercise workload, respectively. The FFA levels (µEq/L) of special ginsenoside (GS) fraction-treated group remained significantly higher at around 880.3 (at rest), around 352.6 (30% maximal oxygen uptake (30% VO_{2max})), around 253.0 (70% maximal oxygen uptake (70% VO_{2max})), around 224.1 (exhaustion) during exercise workload, respectively, when compared to those (4 each maximal oxygen uptake levels) of sterile saline (SAL)-treated group.

Third, no significant differences were observed in changes in liver and muscle (soleus and gastrocnemius) glycogen and lactate (**40**) concentrations after an exhaustive exercise.

Fourth, on the effects of two special active components ginsenoside Rb1 (**6**) and ginsenoside Rg1 (**39**) in special ginsenoside (GS) fraction-mediated enhancement in aerobic exercise performance (aerobic endurance), *first*, the running time (min) regarding the effect of intraperitoneally (*i.p.*) acute injection on exercise endurance of non-trained rats was around 35 (SAL. 1 mL/kg), around 36 (ginsenoside Rb1 (**6**). 2.5 mg/kg), around 37.5 (ginsenoside Rb1 (**6**). 5.0 mg/kg), and around 27.5 (SAL. 1 mg/kg), around 30 (ginsenoside Rg1 (**39**). 2.5 mg/kg), around 29 (ginsenoside Rg1 (**39**). 5.0 mg/kg), respectively.

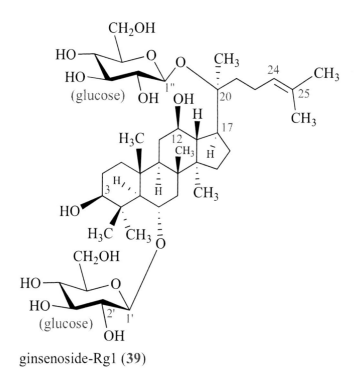

Figure 12. *Panaxatriol Group:* ginsenoside-Rg1 (**39**).

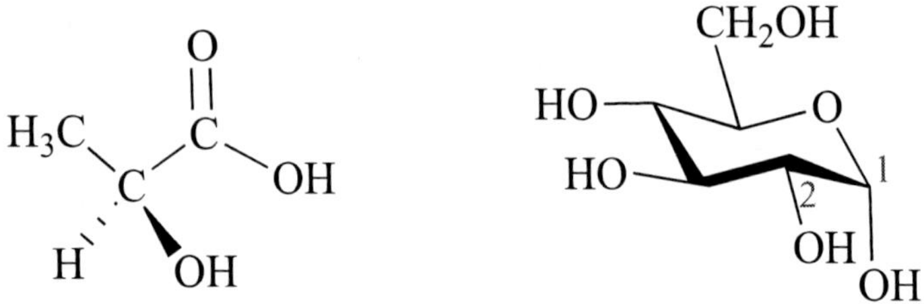

lactate (L-lactic acid. 40)   glucose (41)

Figure 13. Lactate (L-lactic acid. 40) and glucose (41).

Second, the running time (min) regarding the effect of intraperitoneally (*i.p.*) chronic (4 days) injection on exercise endurance of non-trained rats was around 32 (SAL. 1 mL/kg/day), around 50 (ginsenoside Rb1 (**6**). 2.5 mg/kg/day), and around 34 (SAL. 1 mL/kg/day) and around 52 (ginsenoside Rg1 (**39**). 2.5 mg/kg/day), respectively.

From these results (First - Fourth) above, a 4-day specific ginseng root saponin fraction (ginsenoside (GS) fraction) repetitive administration using non-trained adult male Sprague-Dawley (SD) rats extended aerobic endurance significantly, when exercising at around 70% maximal oxygen uptakes (70% VO_{2max}). The ergogenic effect of ginsenoside (GS) fraction could be due to enhancement of plasma free fatty acids (FFA) utilization and sparing use of liver and muscle glycogen. Moreover, it was suggested that either purified ginsenoside Rb1 (**6**) and/or purified ginsenoside Rg1 (**39**) from special ginsenoside (GS) fraction was essential for eliciting the beneficial ergogenic effect of special ginsenoside (GS) fraction [43].

2.1.9.3.2. Effect on Exercise Performance of Ginseng Ginsenosides (Steroid-Like Phytochemicals) in Volunteers

In 2011, a group from School of Nutrition and Health Sciences, Taipei Medical University, Taipei, Taiwan examined the effects of soybean peptides, taurine, Pueraria isoflavone, and ginseng saponin complex drug (four STPG capsules 2 g/day) on exercise performance and endurance exercise capacity in humans, in order to further improve the effect of *Panax ginseng*, by using a 75% maximal oxygen uptake (75% VO_{2max}) exhaustive cycling test (75% maximal oxygen consumption test) together with ergometer for measuring the correlation of exercise and exhaustion. The STPG composite capsule formulation (500 mg/capsule) was combinated by each 4 characteristic phytochemicals - 150 mg soybean peptides, 50 mg taurine, 45 mg Pueraria Radix extract (Pueraria isoflavone), and 30 mg ginseng saponin.

2 capsules (each 500 mg/capsule) of STPG or placebo were orally administrated after breakfast, and after dinner twice per day for 15 days, respectively.

Soybean peptide displayed a role to substantially accelerate metabolism [44, 45] and enhanced exercise performance [46]. A sulfonic amino acid taurine (**42**) (Figure 14) found primarily in skeletal muscle [47, 48] reduced physical fatigue and muscle damage during the exercise training in rats [49]. An antioxidative herbal medicine Pueraria Radix (root of kudzu, *Pueraria lobata*) protects against vascular endothelial cell apoptosis (hypoxia of ischemia induced) by treating endothelial cells with chemical sodium cyanide (NaCN) in glucose-free medium [50]. Ginseng saponin treatment significantly increased the levels of the nonesterified fatty acids (NEFAs), maintained plasma glucose level during exercise, and enhanced aerobic exercise performance [51].

The exercise performance time, blood metabolites, cardiorespiratory responses and energy substrate utilization were measured in 75% maximal oxygen uptake (75% VO_{2max}) exhaustive cycling test (75% maximal oxygen consumption test) after 15-day supplementation.

First, on the mean exercise time (minutes) in cycling exercise time until exhaustion by 75% maximal oxygen uptake (75% VO_{2max}) exhaustive cycling test (75% maximal oxygen consumption test) was around 30.99 minutes for STPG composite capsule formulation (500 mg/capsule)-treated group which was higher when compared to around 28.05 minutes for placebo (P treatment)-treated group.

Second, on blood metabolites during 75% maximal oxygen uptake (75% VO_{2max}) exhaustive cycling test (75% maximal oxygen consumption test) following 15 days of STPG capusule-treated group or placebo (P treatment)-treated group were studied. *First*-an enzyme that transfers phosphate group from ATP to creatine (**43**) (Figure 14) plasma creatine kinase (CK) level (U/L) was STPG treatment group (around 311.1) and placebo treatment group (around 272.3) for administration of prior to the exercise (*Pre-Ex*)(resting), STPG treatment group (around 353.9) and P treatment group (around 318.1 for 10 minutes of exercise later (*Ex-10*), STPG treatment group (around 375.5) and P treatment group (around 331.0) for 15 minutes of exercise later (*Ex-15*), STPG treatment group (around 381.5) and P treatment group (around 324.0) for 20 minutes of exercise later (*Ex-20*), STPG treatment group (around 373.5) and P treatment group (around 270.3) for 25 minutes of exercise later (*Ex-25*), and STPG treatment group (around 381.3) and P treatment group (around 342.1) for exhaustion after exercise (*Exhaustion*), respectively.

Second, plasma ammonia (**44**) (Figure 14) level (μmol/L) was STPG treatment group (around 8.2) and placebo treatment group (around 10.4) for administration of prior to the exercise (*Pre-Ex*), STPG treatment group (around 48.6) and P treatment group (around around 50.4 for 10 minutes of exercise later (*Ex-10*), STPG treatment group (around 84.0) and P treatment group (87.9) for 15 minutes of exercise later (*Ex-15*), STPG treatment group (around 93.3) and P treatment group (around 108.5) for 20 minutes of exercise later (*Ex-20*), STPG treatment group (around 108.0) and P treatment group (around 129.9) for 25 minutes of exercise later (*Ex-25*), and STPG treatment group (around 109.7) and P treatment group (around 118.4) for exhaustion after exercise (*Exhaustion*), respectively.

Third, plasma glucose (**41**) level (mg/dL) was STPG treatment group (around 83.0) and placebo treatment group (around 82.5) for administration of prior to exercise (*Pre-Ex*), STPG treatment group (around 82.6) and P treatment group (around 83.1 for 10 minutes of exercise later (*Ex-10*), STPG treatment group (around 85.9) and P treatment group (85.7) for 15 minutes of exercise later (*Ex-15*), STPG treatment group (around 94.1) and P treatment group (around 90.6) for 20 minutes of exercise later (*Ex-20*), STPG treatment group (around 95.0) and P treatment group (around 102.3) for 25 minutes of exercise later (*Ex-25*), and STPG treatment group (around 114.6) and P treatment group (around 109.4) for exhaustion after exercise (*Exhaustion*), respectively.

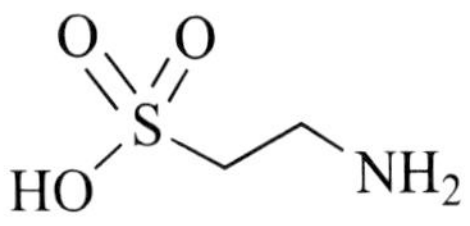

Medicinal Phytochemicals and Health Effects of Panax ginseng ... 63

taurine (42) 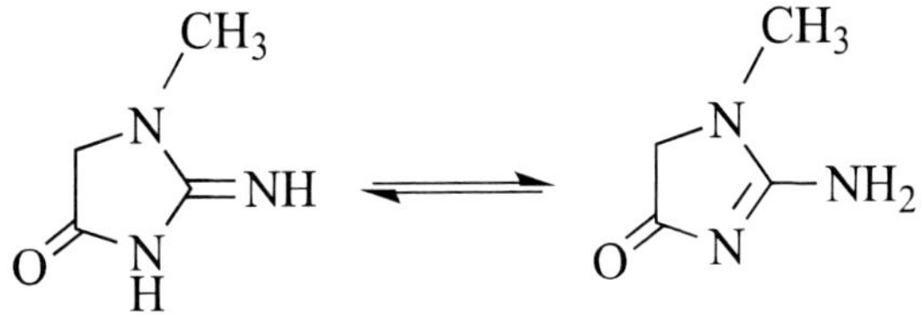   creatine (43)

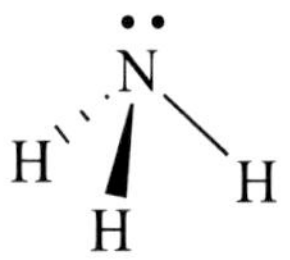

ammonia (44) 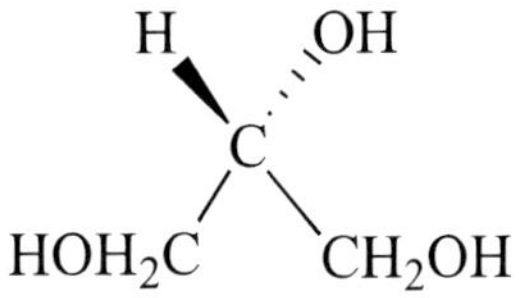   glycerol (glycerin. 45)

Figure 14. Taurine (**42**), creatine (**43**), ammonia (**44**), and glycerol (glycerin. **45**).

Fourth, the plasma glycerol (glycerin. **45**) (Figure 14) level (μmol/L) was STPG treatment group (around 60.0) and placebo treatment group (around 60.6) for administration of prior to exercise (*Pre-Ex*), STPG treatment group (around 101.4) and P treatment group (around 110.7) for 10 minutes of exercise later (*Ex-10*), STPG treatment group (around 128.7) and P treatment group (136.6) for 15 minutes of exercise later (*Ex-15*), STPG treatment group (around 170.1) and P treatment group (around 173.4) for 20 minutes of exercise later (*Ex-20*), STPG treatment group (around 181.2) and P treatment group (around 182.7) for 25 minutes of exercise later (*Ex-25*), and STPG treatment group (around 216.1) and P treatment group (around 215.5) for exhaustion after exercise (*Exhaustion*), respectively.

Fifth, plasma lactate (**40**) level (mmol/L) was STPG treatment group (around 1) and placebo treatment group (around 1) for administration of prior to exercise (*Pre-Ex*), STPG treatment group (around 7.7) and P treatment group (around 8.2) for 10 minutes of exercise later (*Ex-10*), STPG treatment group (around 9.8) and P treatment group (10.5) for 15 minutes of exercise later (*Ex-15*), STPG treatment group (around 10.9) and P treatment group (around 11.5) for 20 minutes of exercise later (*Ex-20*),

STPG treatment (around 11.0) and P treatment group (around 12.0) for 25 minutes of exercise later (*Ex-25*), and STPG treatment group (around 12.0) and P treatment group (around 12.0) for exhaustion after exercise (*Exhaustion*), respectively.

Sixth, plasma non-esterified fatty acids (NEFAs. free fatty acids) level (mmol/L) was STPG treatment group (around 0.35) and placebo treatment group (around 0.36) for administration of prior to exercise (*Pre-Ex*), STPG treatment group (around 0.29) and P treatment group (around 0.24) for 10 minutes of exercise later (*Ex-10*), STPG treatment group (around 0.27) and P treatment group (0.21) for 15 minutes of exercise later (*Ex-15*), STPG treatment group (around 0.325) and P treatment group (around 0.225) for 20 minutes of exercise later (*Ex-20*), STPG treatment group (around 0.326) and P treatment group (around 0.24) for 25 minutes of exercise later (*Ex-25*), and STPG treatment group (around 0.375) and P treatment group (around 0.252) for exhaustion after exercise (*Exhaustion*), respectively.

From the results (First - Sixth) above, in both STPG group and P treatments group, plasma lactate (**40**), ammonia (**44**), and glycerol (glycerin. **45**) significantly increased with the duration of cycling exercise and reached a peak at exhaustion.

Plasma lactate (**40**) levels (mmol/L) at *Ex-20* (around 10.9) and *Ex-25* (around 11.0) in STPG treatment group were lower when compared to those (plasma lactate (**40**) concentrations) of *Ex-20* (around 11.5) and *Ex-25* (around 12.0) in P treatment group.

Additionally, plasma non-esterified fatty acids (NEFAs. free fatty acids) level (mmol/L) at *Ex-15* (around 0.27), *Ex-20* (around 0.325), *Ex-25* (around 0.326), and *exhaustion* (around 0.375), respectively in the STPG treatment group were significantly higher when compared to those (plasma NEFAs) at *Ex-15* (0.21), *Ex-20* (around 0.225), *Ex-25* (around 0.24), and *exhaustion* (around 0.252), respectively in P treatment group.

There were no significant differences in plasma creatine kinase (CK) activities, glucose (**41**) concentrations, or liver function marker enzyme, glutamic oxaloacetic transaminase (GOT) or glutamic pyruvic-transaminase (GPT) activities between STPG treatment group and P treatment treatments for entire experimental period.

Third, on cardiorespiratory responses (heart rate responses) during 75% maximal oxygen uptake (75% VO_{2max}) exhaustive cycling test (75% maximal oxygen consumption test) following 15 days of STPG composite capsule formulation (500 mg/capsule) or placebo (P treatment) administration, *first*, the heart rate (beats/min) was STPG treatment group (around 64.79) and placebo treatment group (around 65.14) for administration of prior to exercise (*Pre-Ex*), STPG treatment group (around 159.64) and P treatment group (around 160.71) for 5 minutes of exercise later (*Ex-5*), STPG treatment group (around 170.29) and P treatment group (around 170.50) for 10 minutes of exercise later (*Ex-10*), STPG treatment group (around 176.21) and P treatment group (around 176.50) for 15 minutes of exercise later (*Ex-15*), STPG treatment group (around 180.00) and P treatment group (around 180.00) for 20 minutes of exercise later (*Ex-20*), STPG treatment group (around 180.50) and P treatment group (around 183.20) for 25 minutes of exercise later (*Ex-25*), and STPG treatment group (around 183.79) and P treatment group (around 183.07) for exhaustion after exercise (*Exhaustion*), respectively.

Second, oxygen consumption (VO_2) (L/min) was STPG treatment group (around 0.31) and placebo treatment group (around 0.31) for administration of prior to exercise (*Pre-Ex*), STPG treatment group (around 2.56) and P treatment group (around 2.38) for 5 minutes of exercise later (*Ex-5*), STPG treatment (around 2.54) and P treatment group (around 2.48) for 10 minutes of exercise later (*Ex-10*), STPG treatment group (around 2.76) and P treatment group (around 2.63) for 15 minutes of exercise later (*Ex-15*), STPG treatment group (around 2.80) and P treatment group (around 2.62) for 20 minutes of exercise later (*Ex-20*), STPG treatment group (around 2.82) and P treatment group (around 2.74) for 25 minutes of exercise later (*Ex-25*), and STPG treatment group (around 2.90) and P treatment group (around 2.74) for exhaustion after exercise (*Exhaustion*), respectively.

Third-carbon dioxide production (VCO_2) levels (L/min) was STPG treatment group (around 0.25) and placebo treatment group (around 0.25) for administration of prior to exercise (*Pre-Ex*), STPG treatment group (around 2.58) and P treatment group (around 2.38) for 5 minutes of exercise later (*Ex-5*), STPG treatment group (around 2.52) and P treatment group (around 2.49) for 10 minutes of exercise later (*Ex-10*), STPG treatment group (around 2.69) and P treatment group (around 2.58) for 15 minutes of exercise later (*Ex-15*), STPG treatment group (around 2.60) and P treatment group (around 2.50) for 20 minutes of exercise later (*Ex-20*), STPG treatment group (around 2.47) and P treatment group (around 2.40) for 25 minutes of exercise later (*Ex-25*), and STPG treatment group (around 2.42) and P treatment group (around 2.38) for exhaustion after exercise (*Exhaustion*), respectively.

Fourth, respiratory exchange ratio (RER) level was STPG treatment group (around 0.81) and placebo treatment group (around 0.82) for administration of prior to exercise (*Pre-Ex*), STPG treatment group (v1.01) and P treatment group (around 1.00) for 5 minutes of exercise later (*Ex-5*), STPG treatment group (around 0.99) and P treatment group (around 1.01) for 10 minutes of exercise later (*Ex-10*), STPG treatment group (around 0.98) and P treatment group (v0.98) for 15 minutes of exercise later (*Ex-15*), STPG treatment group (v0.94) and P treatment group (v0.96) for 20 minutes of exercise later (*Ex-20*), STPG treatment group (around 0.88) and P treatment group (around 0.88) for 25 minutes of exercise later (*Ex-25*), and STPG treatment group (around 0.83) and P treatment group (around 0.87) for exhaustion after exercise (*Exhaustion*), respectively.

From the results above, heart rate in both STPG treatment group and P treatment group increased more with exercise performance time when compared to that (heart rate) of prior to exercise (*Pre-Ex*) and reached up to around 3 times at a peak of exhaustion. However, there were no significant differences between STPG treatment group and P treatment group.

The oxygen consumption (VO_2) levels (L/min) at *Ex-5* (around 2.56), *Ex-20* (around 2.80), and *Exhaustion* (around 2.90) in STPG treatment group were significantly higher when compared to those (VO_2 levels (L/min)) of *Ex-5* (around 2.38), *Ex-20* (around 2.62), and *Exhaustion* (around 2.74) in P treatment group.

The carbon dioxide production (VCO_2) levels (L/min) at *Ex-5* (around 2.58) and *Ex-15* (around 2.69) in STPG treatment group also were significantly higher when compared to those (VCO_2 levels (L/min)) of *Ex-5* (around 2.38) and *Ex-15* (around 2.58) in P treatment group.

Additionally, respiratory exchange ratio (RER) levels at all seven time points (prior to exercise (*Pre-Ex*), 5 minutes of exercise later (*Ex-5*), 10 minutes of exercise later (*Ex-10*), 15 minutes of exercise later (*Ex-15*), 20 minutes of exercise later (*Ex-20*), 25 minutes of exercise later (*Ex-25*), and exhaustion after exercise (*Exhaustion*) during exercise performance time were more prominent when compared to those (RER levels) at prior to exercise (*Pre-Ex*), and progressively decreased from 5 minutes of exercise later (*Ex-5*) to exhaustion after exercise (*Exhaustion*). However, there were no differences between STPG treatment group and P treatment group at all same six exercise time points.

Fourth, on whole-body carbohydrate oxidation (utilization of carbohydrate) and whole-body fat oxidation during 75% maximal oxygen uptake (75% VO_{2max}) exhaustive cycling test (75% maximal oxygen consumption test) following 15 days of STPG consumption or P consumption (n=14 per treatment), *first*-whole-body carbohydrate oxidation (utilization of carbohydrate) level (L/min) was STPG treatment group (around 3.55) and P treatment group (around 3.3) for 5 minutes of exercise later (*Ex-5*), STPG treatment group (around 3.4) and P treatment group (around 3.4) for 10 minutes of exercise later (*Ex-10*), STPG treatment group (around 3.45) and P treatment group (around 3.3) for 15 minutes of exercise later (*Ex-15*), STPG treatment group (around 2.90) and P treatment group (around 3.0) for 20 minutes of exercise later (*Ex-20*), STPG treatment group (around 2.25) and P treatment group (around 2.10) for 25 minutes of exercise later (*Ex-25*), and STPG treatment group (around 1.7) and P treatment group (around 2.0) for exhaustion after exercise (*Exhaustion*), respectively.

Second, whole-body fat oxidation level (L/min) was STPG treatment group (around 0.02) and P treatment group (around 0.02) for 5 minutes of exercise later (*Ex-5*), STPG treatment group (around 0.03) and P treatment group (around 0.05) for 10 minutes of exercise later (*Ex-10*), STPG treatment group (around 0.17) and P treatment group (around 0.15) for 15 minutes of exercise later (*Ex-15*), STPG treatment group (around 0.46) and

P treatment group (around 0.25) for 20 minutes of exercise later (*Ex-20*), STPG treatment group (around 0.59) and P treatment group (around 0.57) for 25 minutes of exercise later (*Ex-25*), and STPG treatment group (around 0.80) and P treatment group (around 0.60) for exhaustion after exercise (*Exhaustion*), respectively. From the results *(First* and *Second)* above, there were no differences between STPG treatment group and P treatment group at all same six exercise time points.

From these results (First - Fourth), respiratory metabolic analysis showed that STPG composite capsule formulation (500 mg/capsule) increased lipid utilization in the body as an energy substrate when compared to that (lipid utilization in the body) of P (placebo) during 75% maximal oxygen uptake (75% VO_{2max}) exhaustive cycling test (75% maximal oxygen consumption test). Additionally, two blood lactate (**40**) levels (mmol/L) - around 10.9 for 20 minutes of exercise later (*Ex-20*) and around 11.0 for 25 minutes of exercise later (*Ex-25*) - in the STPG treatment group was significantly lower when compared to that of *Ex-20* (around 11.5) and *Ex-25* (around 12.0) in P treatment group during 75% maximal oxygen uptake (75% VO_{2max}) exhaustive cycling test (75% maximal oxygen consumption test).

The lipid metabolism utilization in body as energy substrate during exercise can facilitate aerobic exercise capacity (aerobic performance) because obtaining energy from carbohydrates leads to lactate (**40**) production and a pH decrease that could inhibit muscle contractions [52, 53, 54, 55]. Additionally, the depletion of carbohydrate results in deposition of glycogen, making it difficult to continue exercise. Therefore, acceleration of lipid metabolism would lead to improved endurance.

Moreover, it is known that facilitation of lipid utilization by muscles is more effective when compared to lipid metabolism in adipose tissues during exercise.

Regulation of fatty acyl-coenzyme A entry into mitochondria by a mitochondrial outer membrane enzyme catalyzing binding of fatty acid and carnitine palmitoyltransferase-I (CPT-1) is a rate-limiting step in the oxidation of fatty acids in muscles [56].

Actually, plasma non-esterified fatty acids (NEFAs. free fatty acids) levels (mmol/L) at *Ex-15* (around 0.27), *Ex-20* (around 0.325), *Ex-25* (around 0.326), and *exhaustion* (around 0.375) in STPG treatment group were higher when compared to those (*Ex-15* (0.21), *Ex-20* (around 0.225), *Ex-25* (around 0.24), and *exhaustion* (around 0.252)) in P treatment group. The enhancement of lipid metabolism may be partly due to the antioxidative effect of Pueraria Radix (dried kuduz root, *Pueraria lobata*).

A short-term (4-day) treatment with ginseng saponin (GS) (10 and 20 mg/kg/day) significantly prolonged aerobic endurance of non-trained rats exercising at around 70% maximal oxygen uptake (70% VO_{2max}) exhaustive cycling test (70% maximal oxygen consumption test). Ginseng saponin (GS) treatment group significantly increased plasma free fatty acid (FFA) level when compared to that (increase of plasma FFA level) of a saline control group and maintained plasma glucose level during exercise [57]. Therefore, it suggested that ginseng saponins (GS) contributed to increase in exercise endurance by the administration of STPG composite capsule formulation (500 mg/capsule).

Furthermore, it suggested that administration of 4 STPG capsules (total 2 g) per day (500 mg/capsule) for 15 days increased the utilization of plasma free fatty acid (FFA) as energy source by sparing of glycogen during prolonged exercise, and that STPG composite capsule formulation could apparently enhance the endurance performance for athletes and exercising individuals including a long endurance event [58].

2.1.10. Immunomodulating Activities of Polysaccharides

It is known that ginseng root polysaccharides have mitogenic [59], hypoglycemic [60], antitumor [61], and macrophage cytokine-inducing [62] activities.

Ginseng marc is a fibrous and insoluble by-product remaining after the extraction process of ginseng. Based on the above facts, in 2004, a group from Department of Biotechnology & Bioproducts Research Center, Yonsei University, Seoul, Korea examined in *vitro* immunomodulatory effects of polysaccharides isolated from extruded ginseng marc, using murine peritoneal macrophages.

First, regarding the effect of ginseng marc polysaccharide (GMP) on lysosomal enzyme activity of peritoneal macrophages, macrophages were treated with doses of 50 and 100 μg/mL ginseng marc polysaccharide (GMP) and 20 μg/mL lipopolysaccharide (LPS) from *Escherichia coli* serotype for 48 hrs. The lysosomal phosphatase activity (%) was 100 for a control 0 μg/mL, ginseng marc polysaccharide (GMP) around 142 for 50 μg/mL GMP, around 160 for 100 μg/mL GMP and a positive control around 225 for 20 μg/mL LPS, respectively.

From the results, among 4 samples (a control 0 μg/mL GMP, 50 μg/mL GMP, 100 μg/mL GMP, and a positive control 20 μg/mL LPS), lysosomal phosphatase activity (%) stimulated by 20 μg/mL LPS was significantly 220% higher when compared to that (lysosomal phosphatase activity) of a positive control 20 μg/mL LPS. The lysosomal enzyme activity of peritoneal macrophages stimulated by ginseng marc polysaccharide (GMP) was significantly increased, by 142% at 50 μg/mL GMP, and 160% at 100 μg/mL GMP, respectively, when compared to that (lysosomal phosphatase activity (%)) of a control 0 μg/mL GMP.

Second, regarding the effect of GMP on phagocytic index (P.I.) (%) of peritoneal macrophages, phagocytic index (P.I.) (%) was 100 for a control 0 μg/mL ginseng marc polysaccharide (GMP), around 232 for 100 μg/mL GMP, and around 348 for a positive control 20 μg/mL LPS, respectively.

From these results, among 3 samples (a control 0 μg/mL GMP, 100 μg/mL GMP, and a positive control 20 μg/mL LPS), the phagocytic index (P.I.) (%) of phagocytosis stimulated by 20 μg/mL LPS was 348% higher when compared to that (phagocytic index (P.I.) (100%)) of a control 0 μg/mL GMP. Additionally, phagocytic index (P.I.) (%) of peritoneal macrophages treated with 100 μg/mL GM was 232% higher when compared to that (phagocytic index (P.I.) (100%)) of a control 0 μg/mL GMP.

Third, regarding the effect of GMP on nitrite (NO) production of peritoneal macrophages, production of nitrile (%) was 100 for a control 0 μg/mL ginseng marc polysaccharide (GMP), around 244 for 50 μg/mL

GMP, around 269 for 100 μg/mL GMP, and a positive control around 408 for 20 μg/mL LPS, respectively.

From these results, among 4 samples (a control 0 μg/mL GMP, 50 μg/mL GMP, 100 μg/mL GMP, and a positive control 20 μg/mL LPS), the effect on nitrite production of GMP as peritoneal macrophage function was also assessed through changes in the release of two reactive oxygen intermediates (ROIs) (NO and H_2O_2). The nitrite production by peritoneal macrophages stimulated by 20 μg/mL LPS was 408% higher when compared to that (nitrite production. 100%) of a control 0 μg/mL GMP. The administration of GMP significantly increased nitrite (NO) production of macrophages, by 244% at 50 μg/mL GMP and 269% at 100 μg/mL GMP, respectively, when compared to that (nitrite production. 100%) of a control 0 μg/mL GMP.

The increase in nitrite (NO) production by GMPs is also important information because nitrite (NO) release by peritoneal macrophages has been known to be the predominant mechanism by which infectious agents are destroyed [63]. Additionally, GMP induced a dose-related increase in hydrogen peroxide (H_2O_2) by peritoneal macrophages. The ability of GMP to induce hydrogen peroxide (H_2O_2) might contribute to ability of ginseng to produce early immune response mediators and to have antiviral effects. These immunomodulatory properties of GMP could be associated with its claimed prophylactic and therapeutic effects against common cold [64].

Fourth, regarding the effect of GMP on hydrogen peroxide (H_2O_2) production of peritoneal macrophages, the production of hydrogen peroxide (H_2O_2) (%) was 100 for a control 0 μg/mL ginseng marc polysaccharide (GMP), around 131 for 50 μg/mL GMP, around 139 for 100 μg/mL GMP, and a positive control around 167 for 20 μg/mL LPS, respectively.

From the results, among 4 samples (a control 0 μg/mL GMP, 50 μg/mL GMP, 100 μg/mL GMP, and a positive control 20 μg/mL LPS), the effect on production of hydrogen peroxide (H_2O_2) (%), the production of hydrogen peroxide (H_2O_2) (%) by peritoneal macrophages stimulated by 20 μg/mL LPS was 167% higher when compared to that (hydrogen peroxide

(H₂O₂) production (100%)) of a control 0 µg/mL GMP. GMP significantly increased hydrogen peroxide (H₂O₂) production (%) of peritoneal macrophages, by 131% at 50 µg/mL GMP and 139% at 100 µg/mL GMP, respectively, when compared to that (hydrogen peroxide (H₂O₂) production (100%)) of control group.

Fifth, regarding the effect of GMP on survival of peritoneal macrophages, the survival effect (%) was 100 for a control 0 µg/mL ginseng marc polysaccharide (GMP), around 133 for 100 µg/mL GMP, and around 120 for a positive control 20 µg/mL LPS, respectively.

From these results, among 3 samples (a control 0 µg/mL GMP, 100 µg/mL GMP, and a positive control 20 µg/mL LPS), survival of peritoneal macrophages was inhibited in LPS-treated macrophages, at the same time, significantly enhanced in GMP-treated macrophages. The growth index of GMP-treated peritoneal macrophages at 100 µg/mL GMP was higher 133% when compared to that (survival effect (100%)) of a control 0 µg/mL GMP.

From these results (First to Fifth) above, it was demonstrated that GMP resulted in augmentation of peritoneal macrophage function. As a result, GMP could be a strong stimulator for the release of cytotoxic mediators to produce relevant cytokines like lysosomal acid phosphatase (LAP), phagocytic index (P.I.) (%), nitrite (NO) production, hydrogen peroxide (H₂O₂) production, and growth index. Moreover, these results suggested that GMP was one of the effective nonspecific immunomodulatory agents, and that the immunostimulating effects of ginseng polysaccharides (GP) might be due to their stimulation of two reactive oxygen intermediates (ROIs) (NO and H₂O₂)-producing mechanism [65].

2.1.11. Memory Enhancing and Neuroprotective Effects

The wide effects of ginseng extracts on central nervous system (CNS) have been known to improve learning and memory in normal, aged or brain-damaged animals [66, 67, 68]. Ginsenosides are saponin constituents in ginseng, and generally it is thought that these ginsenosides are bioactive components which show diverse beneficial effects such as the enhancement of memory function.

Based on the these facts, in 2005, a group from College of Pharmacy, Chungnam National University, Daejeon, Korea examined the effects of the three ginsenosides - Rg$_3$(R) and Rg$_3$(S), diastereoisomeric forms of ginsenoside Rg$_3$ (**46**) (Figure 15), and Rg$_5$/Rk$_1$, a 1:1 (w/w) mixture of ginsenoside Rg$_5$ (**47**) (Figure 15) and ginsenoside Rk$_1$ (**48**) (Figure 15) - on glutamate (**49**) (Figure 16)- or *N*-methyl-D-aspartate (NMDA. **50**) (Figure 16)-induced excitotoxic damage and on the oxidative stress-induced neuronal damage by using primary cultured rat cortical cells.

First, on the effects of ginsenosides on ethanol-induced memory impairment in Male ICR mice (weighing 18-20 g), the mice were orally administered with saline (vehicle), ginsenoside Rg$_3$(R) (**46 R**) (10 mg/kg), ginsenoside Rg$_3$(S) (**46 S**) (10 mg/kg), or ginsenoside Rg$_5$ (**47**)/ginsenoside Rk$_1$ (**48**) (10 mg/kg) once a day for 4 days, and their memory function was examined by a passive avoidance test of memory acquisition tests after ethanol-induced amnesia. The latency learning is a form of learning that is not immediately expressed in overt response.

First-the latency learning time (latency learning period) (seconds) of cognitive performance in amnesia mice was around 30 seconds for vehicle (saline) with no ethanol treatment (-), around 30 sec. for vehicle (saline) with ethanol treatment (+), around 40 sec. for ginsenoside Rg$_3$(R) (**46 R**) (10 mg/kg) with ethanol treatment (+), around 35 sec. for ginsenoside Rg$_3$(S) (**46 S**) (10 mg/kg) with ethanol treatment (+), and around 45 sec. for ginsenoside Rg$_5$ (**47**)/ginsenoside Rk$_1$ (**48**) (10 mg/kg) with ethanol treatment (+), respectively.

Second-the latency testing time (latency testing period) (seconds) of cognitive performance in amnesia mice was around 175 seconds for vehicle (saline) with no ethanol treatment (-), around 45 sec. for vehicle (saline) with ethanol treatment (+), around 125 sec. for ginsenoside Rg$_3$(R) (**46 R**) (10 mg/kg) with ethanol treatment (+), around 115 sec. for ginsenoside Rg$_3$(S) (**46 S**) (10 mg/kg) with ethanol treatment (+), and around 210 sec. for ginsenoside Rg$_5$ (**47**)/ginsenoside Rk$_1$ (**48**) (10 mg/kg) with ethanol treatment (+), respectively.

These results suggested, the latency testing time (latency testing period) of ethanol-treated mice was significantly increased by oral administration of ginsenoside Rg$_3$(R) (**46** R), ginsenoside Rg$_3$(S) (**46** S) or 1:1 mixture ginsenoside Rg$_5$ (**47**)/ginsenoside Rk$_1$ (**48**) for 4 days by using Mann-Whitney Utest. Although all three ginsenosides were effective, ginsenoside Rg$_5$ (**47**)/ginsenoside Rk$_1$ (**48**) showed utmost effect for reversing the shorting of latency testing time (latency testing period). The latency testing period (190 seconds) of ginsenoside Rg$_5$ (**47**)/ginsenoside Rk$_1$ (**48**)-treated mice was around 4.7 times longer latency period when compared to that (latency learning period) (80 seconds) of ginsenoside Rg$_5$ (**47**)/ginsenoside Rk$_1$ (**48**) as the control mice. Each latency period (125 sec. and 115 sec.) of both ginsenoside Rg$_3$(R) (**46** R)- and ginsenoside Rg$_3$(S) (**46** S)-treated mice showed about around 3.2 and 3.3 times longer latency period, respectively, when compared to the corresponding latency learning period (40 sec. and 35 sec.) as ethanol-treated control mice.

Second, on the effects of ginsenosides on scopolamine (hyoscine. **51**) (Figure 16) -induced memory impairment in Male ICR mice (weighing 18-20 g), the mice were orally administered with vehicle (saline), ginsenoside Rg$_3$(R) (**46** R) (10 mg/kg), ginsenoside Rg$_3$(S) (**46** S) (10 mg/kg), or ginsenoside Rg$_5$ (**47**)/ginsenoside Rk$_1$ (**48**) (10 mg/kg) once a day for 4 days, and their memory function was examined by a passive avoidance test of memory acquisition tests after scopolamine (hyoscine. **51**)-induced amnesia. The memory impairment of mice was induced by single intraperitoneal (*i.p.*) injection of scopolamine (hyoscine. **51**) (3 mg/kg), and the effects of ginsenosides were examined on memory impairment.

First-the latency learning time (latency learning period) (seconds) of cognitive performance in amnesia mice was around 55 seconds for vehicle (saline) with no scopolamine (hyoscine. **51**) (-), around 50 sec. for vehicle (saline) with scopolamine (hyoscine. **51**) treatment (+), around 70 sec. for ginsenoside Rg$_3$(R) (**46** R) (10 mg/kg) with scopolamine (hyoscine. **51**) treatment (+), around 55 sec. for ginsenoside Rg$_3$(S) (**46** S) (10 mg/kg) with scopolamine (hyoscine. **51**) treatment (+) and around 80 sec. for ginsenoside Rg$_5$ (**47**)/ginsenoside Rk$_1$ (**48**) (10 mg/kg) with scopolamine (hyoscine. **51**) treatment (+), respectively.

Second-the latency testing time (latency testing period) (seconds) of cognitive performance in amnesia mice was around 165 seconds for vehicle (saline) with no scopolamine (hyoscine. **51**) treatment (-), around 40 sec. for vehicle (saline) with scopolamine (hyoscine. **51**) treatment (+), around 60 sec. for ginsenoside Rg$_3$(*R*) (**46** *R*) (10 mg/kg) with scopolamine (hyoscine. **51**) treatment (+), around 100 sec. for ginsenoside Rg$_3$(*S*) (**46** *S*) (10 mg/kg) with scopolamine (hyoscine. **51**) treatment (+), and around 190 sec. for ginsenoside Rg$_5$ (**47**)/ginsenoside Rk$_1$ (**48**) (10 mg/kg) with scopolamine (hyoscine. **51**) treatment (+), respectively.

From these results, scopolamine (hyoscine. **51**) showed similar degree of memory impairment as the ethanol group. The latency testing period of scopolamine (hyoscine. **51**)-treated mice was significantly increased by oral administration of ginsenoside Rg$_3$(*S*) (**46** *S*) or ginsenoside Rg$_5$ (**47**)/ginsenoside Rk$_1$ (**48**), from Mann-Whitney U-test. The combination (1:1 mixture) of ginsenoside Rg$_5$ (**47**)/ginsenoside Rk$_1$ (**48**) exhibited the most effective reversal for shorting of latency testing period among three ginsenosides – ginsenoside Rg$_3$(*R*) (**46** *R*), ginsenoside Rg$_3$(*S*) (**46** *S*), and ginsenoside Rg$_5$ (**47**)/ginsenoside Rk$_1$ (**48**). The latency testing period in scopolamine (hyoscine. **51**)-treated mice was increased by ginsenoside Rg$_5$ (**47**)/ginsenoside Rk$_1$ (**48**) up to around 2.4-fold when compared to the latency learning period of ginsenoside Rg$_5$ (**47**)/ginsenoside Rk$_1$ (**48**) as control mice. Similarly, although ginsenoside Rg$_3$(*S*) (**46** *S*) showed similar trend effect in administration of both ethanol-induced and scopolamine (hyoscine. **51**)-induced amnesia models, ginsenoside Rg$_3$(*R*) (**46** *R*) did not show any increase in the latency testing period in scopolamine (hyoscine. **51**)-induced amnesia group.

Third, on the effects of three ginsenosides – ginsenoside Rg$_3$(*R*) (**46** *R*), ginsenoside Rg$_3$(*S*) (**46** *S*), ginsenoside Rg$_5$ (**47**)/ginsenoside Rk$_1$ (**48**) - on excitotoxicity induced by glutamate (glutamic acid. **49**) or *N*-methyl-D-aspartate (NMDA. **50**) in primary cultured cortical cells, the primary cultured cortical cells (cultured 10-12 days *in vitro*) were exposed to 100 μM glutamate (glutamic acid. **49**) or *N*-methyl-D-aspartate (NMDA. **50**) for 15 min in the absence or presence of indicated ginsenoside concentrations. The cell damage such as excitotoxic neuronal cell damage or oxidative neuronal cell damage was assessed after 20-24 hrs by measuring lactate dehydrogenase (LDH. lactic acid dehydrogenase) activity in culture media. Here, lactate dehydrogenase (LDH. lactic acid dehydrogenase) is one of the group of enzymes found in blood and other body tissues which is involved in energy production in cells. The increased amount of lactate dehydrogenase (LDH. lactic acid dehydrogenase) in blood might be a sign of tissue damage and some types of cancer or other diseases.

First-LDH (lactate dehydrogenase. lactic acid dehydrogenase) release (% control) induced by glutamate (glutamic acid. **49**) was around 90 (3 μg/mL), around 100 (10 μg/mL), around 90 (30 μg/mL), around 35 (100 μg/mL), around 2 (300 μg/mL) for ginsenoside Rg$_3$(R) (**46 R**), around 95 (3 μg/mL), around 77 (10 μg/mL), around 10 (30 μg/mL), around 34 (100 μg/mL) for ginsenoside Rg$_3$(S) (**46 S**), and around 90 (3 μg/mL), around 72 (10 μg/mL), around 12 (30 μg/mL), around 18 (100 μg/mL) for ginsenoside Rg$_5$ (**47**)/ginsenoside Rk$_1$ (**48**), respectively.

Second-LDH (lactate dehydrogenase. lactic acid dehydrogenase) release (% control) induced by *N*-methyl-D-aspartate (NMDA. **50**) was around 85 (3 μg/mL), around 75 (10 μg/mL), around 75 (30 μg/mL), around 24 (100 μg/mL), around 6 (300 μg/mL) for ginsenoside Rg$_3$(R) (**46 R**), around 77 (3 μg/mL), around 48 (10 μg/mL), around 8 (30 μg/mL), around 38 (100 μg/mL) for ginsenoside Rg$_3$(S) (**46 S**), and around 70 (3 μg/mL), around 40 (10 μg/mL), around 16 (30 μg/mL), around 8 (100 μg/mL) for ginsenoside Rg$_5$ (**47**)/ginsenoside Rk$_1$ (**48**), respectively.

The effects of three ginsenosides - ginsenoside Rg$_3$(R) (**46 R**), ginsenoside Rg$_3$(S) (**46 S**), ginsenoside Rg$_5$ (**47**)/ginsenoside Rk$_1$ (**48**) - on the cell damage such as excitotoxic neuronal cell damage or oxidative neuronal cell damage, the employed primary cultured rat cortical cells had been maintained for 10-12 days *in vitro*. The excitotoxicity was induced by exposure of primary cultured rat cortical cells for 15 min to either 100 μM glutamate (glutamic acid. **49**) or *N*-methyl-D-aspartate (NMDA. **50**). Prominent neuronal damage was produced after 20-24 hr of exposure to glutamate (glutamic acid. **49**) or *N*-methyl-D-aspartate (NMDA. **50**). By

LDH (lactate dehydrogenase. lactic acid dehydrogenase) measurements, around 70-80% of the cells were damaged compared to vehicle-treated control cells. The excitotoxicity induced by glutamate (glutamic acid. **49**) or *N*-methyl-D-aspartate (NMDA. **50**) was dose-dependently inhibited by three ginsenosides - ginsenoside Rg$_3$(*R*) (**46** *R*), ginsenoside Rg$_3$(*S*) (**46** *S*), ginsenoside Rg$_5$ (**47**)/ginsenoside Rk$_1$ (**48**). Fifty percent inhibitory concentration (IC50) (μg/mL) of glutamate (glutamic acid. **49**)-induced toxicity was achieved at 75.3 (ginsenoside Rg$_3$(*R*) (**46** *R*)), 16.0 (ginsenoside Rg$_3$(*S*) (**46** *S*)), and 14.7 (ginsenoside Rg$_5$ (**47**)/ginsenoside Rk$_1$ (**48**)), respectively. Similarly, 50% inhibitory concentration (IC50) value (μg/mL) for *N*-methyl-D-aspartate (NMDA. **50**)-induced toxicity was achieved at 27.8 (ginsenoside Rg$_3$(*R*) (**46** *R*)), 9.3 (ginsenoside Rg$_3$(*S*) (**46** *S*)), and 6.9 (ginsenoside Rg$_5$ (**47**)/ginsenoside Rk$_1$ (**48**)), respectively. Especially, ginsenoside Rg$_5$ (**47**)/ginsenoside Rk$_1$ (**48**) showed the highest inhibition of excitotoxic neuronal damage.

Fourth, on the repairing effects of three ginsenosides – ginsenoside Rg$_3$ (*R*) (**46** *R*), ginsenoside Rg$_3$(*S*) (**46** *S*), ginsenoside Rg$_5$ (**47**)/ginsenoside Rk$_1$ (**48**) - on the oxidative damage induced by hydrogen peroxide (H$_2$O$_2$) or xanthine/xanthine oxidase system in primary cultured cortical cells, the primary cultured cortical cells were exposed to either 100 μM hydrogen peroxide (H$_2$O$_2$) for 5 min or to xanthine/xanthine oxidase (0.5 mM/10 mU/mL) for 10 min in the absence or presence of tested three ginsenosides.

First-LDH (lactate dehydrogenase. lactic acid dehydrogenase) release (% control) induced by hydrogen peroxide (H$_2$O$_2$) was around 104 (3 μg/mL), around 100 (10 μg/mL), around 100 (30 μg/mL), around 104 (100 μg/mL), around 95 (300 μg/mL) for ginsenoside Rg$_3$(*R*) (**46** *R*), around 98 (3 μg/mL), around 106 (10 μg/mL), around 95 (30 μg/mL), around 88 (100 μg/mL), around 84 (300 μg/mL) for ginsenoside Rg$_3$(*S*) (**46** *S*), and around 106 (3 μg/mL), around 100 (10 μg/mL), around 95 (30 μg/mL), around 82 (100 μg/mL), v86 (300 μg/mL) for ginsenoside Rg$_5$ (**47**)/ginsenoside Rk$_1$ (**48**), respectively.

Second-LDH (lactate dehydrogenase. lactic acid dehydrogenase) release (% control) induced by xanthine/xanthine oxidase was around 102

(3 µg/mL), v100 (10 µg/mL), around 94 (30 µg/mL), around 95 (100 µg/mL), around 87 (300 µg/mL) for ginsenoside Rg₃(R) (**46** R), around 100 (3 µg/mL), around 95 (10 µg/mL), around 85 (30 µg/mL), around 80 (100 µg/mL), around 92 (300 µg/mL) for ginsenoside Rg₃(S) (**46** S), and around 97 (3 µg/mL), around 92 (10 µg/mL), around 95 (30 µg/mL), around 90 (100 µg/mL) and around 92 (300 µg/mL) for ginsenoside Rg₅ (**47**)/ginsenoside Rk₁ (**48**), respectively.

The exposure of primary cultured cortical cells to hydrogen peroxide (H_2O_2) or xanthine/xanthine oxidase produced severe oxidative neuronal cell damage. By LDH (lactate dehydrogenase. lactic acid dehydrogenase) activity measurements, around 80-90% of primary cultured cortical cells were damaged. Hydrogen peroxide (H_2O_2)-induced oxidative damage or xanthine/xanthine oxidase-induced oxidative damage was not effectively inhibited by the ginsenosides that were tested in this study. Additionally, three ginsenosides - ginsenoside Rg₃(R) (**46** R), ginsenoside Rg₃(S) (**46** S), ginsenoside Rg₅ (**47**)/ginsenoside Rk₁ (**48**) - showed no significant effects on 2,2-diphenyl-1-picrylhydrazyl (DPPH. **52**) (Figure 16) radicals. Only ginsenoside Rg₃(S) (**46** S) showed a modest inhibition (22.9 ± 1.4%) on lipid peroxidation in rat brain homogenates induced by Fe^{2+} and ascorbic acid (**53**) (Figure 16) at concentrations of 300 µg/mL.

These results (First to Fourth) above, taken together, indicated that ginsenoside Rg₃(S) (**46** S) and ginsenoside Rg₅ (**47**)/ginsenoside Rk₁ (**48**) significantly reversed memory dysfunction that was induced by ethanol or scopolamine (hyoscine. **51**), and three ginsenosides - ginsenoside Rg₃(R) (**46** R), ginsenoside Rg₃(S) (**46** S), ginsenoside Rg₅ (**47**)/ginsenoside Rk₁ (**48**) - protected primary cultured cortical cells from excitotoxic damage. This is the first report on memory enhancing effects and neuroprotective effects of ginsenoside Rg₅ (**47**)/ginsenoside Rk₁ (**48**) and ginsenoside Rg₃(S) (**46** S). A 1:1 mixture ginsenoside Rg₅ (**47**)/ginsenoside Rk₁ (**48**) was the most effective and potent. Besides, these results suggested that ginsenoside Rg₅ (**47**)/ginsenoside Rk₁ (**48**) and ginsenoside Rg₃(S) (**46** S) could become useful therapeutic potentials for the clinical management of memory loss, observed in aging or neurodegenerative disorders such as Alzheimer's disease (AD) [69].

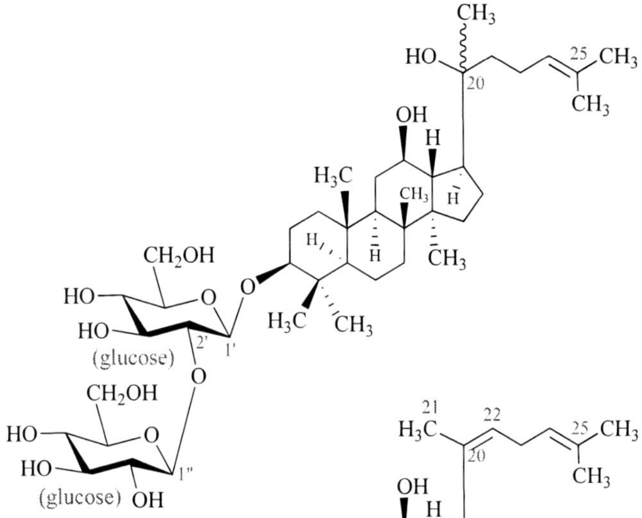

ginsenoside Rg_3 (**46**)

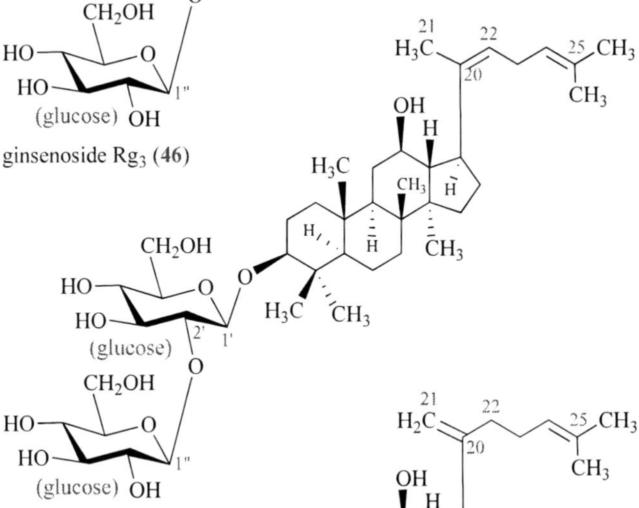

ginsenoside Rg_5 (**47**)

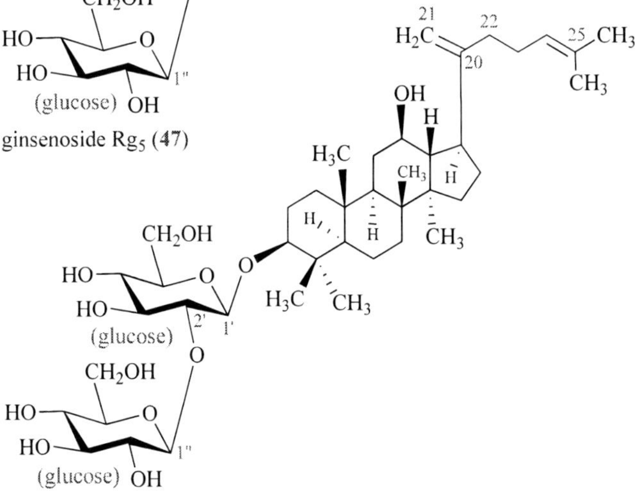

ginsenoside Rk_1 (**48**)

Figure 15. Ginsenoside Rg_3 (**46**), ginsenoside Rg_5 (**47**), and ginsenoside Rk_1 (**48**).

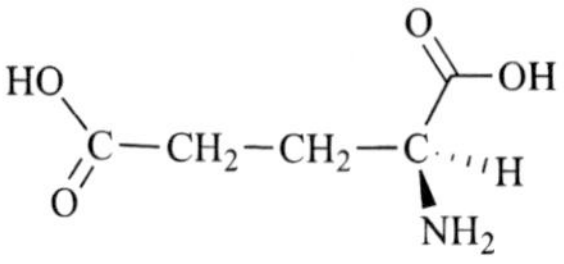

glutamate (glutamic acid. 49)

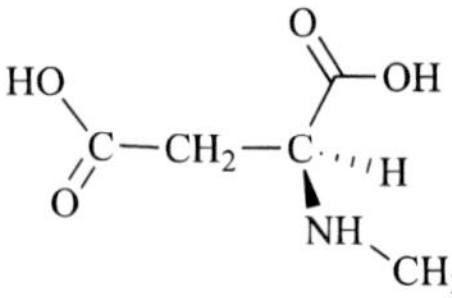

N-methyl-D-aspartate (NMDA. 50)

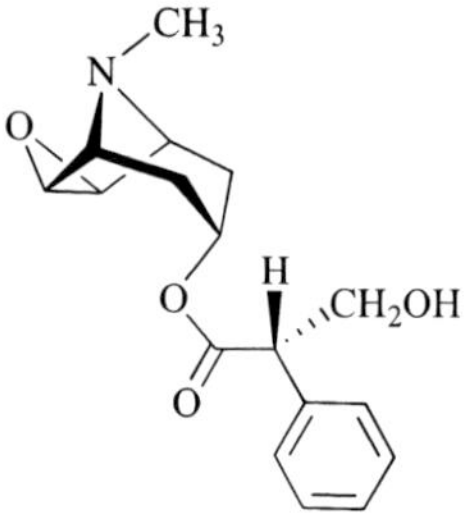

scopolamine (hyoscine. 51)

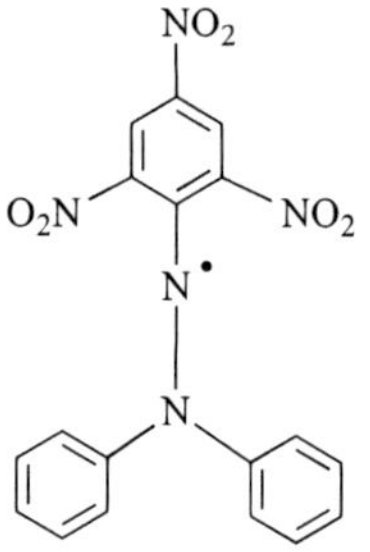

2,2-diphenyl-1-picrylhydrazyl (DPPH. 52)

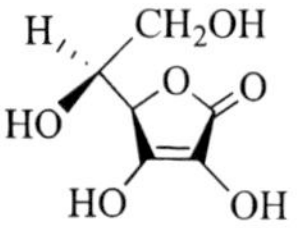

ascorbic acid (53)

Figure 16. Glutamate (glutamic acid. **49**), and *N*-methyl-D-aspartate (NMDA. **50**), scopolamine (hyoscine. **51**), 2,2-diphenyl-1-picrylhydrazyl (DPPH. **52**), and ascorbic acid (**53**).

3. DOSAGE, AND AVOIDANCE FOR ADMINISTRATION

Generally, oral administration dosage of *Panax ginseng* extract is a dosage of around 200 mg/day [70]. The recommended daily dose for *Panax ginseng* is 0.5 to 2 g/day of dried root, or 100 to 300 mg/day of a standardized extract containing 1.5 to 7% ginsenosides. Patients should be cautioned not to exceed these dosages, as adverse reactions such as hypertension, diarrhea, nervousness, and insomnia are associated with excessive ingestion. It is known that there are drug interactions of *Panax ginseng* with both phenelzine (**54**) - a non-selective and irreversible monoamine oxidase inhibitor (MAOI) and warfarin (**55**) (Figure 17) - an anticoagulant (blood thinner). Additionally, the use of *Panax ginseng* is contraindicated for children and women who are pregnant or lactating [70, 71].

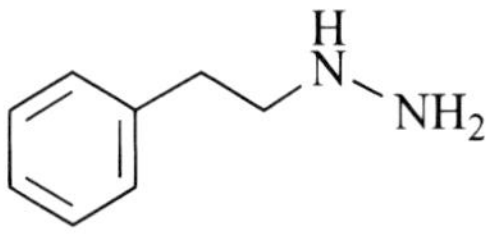

phenelzine (54) 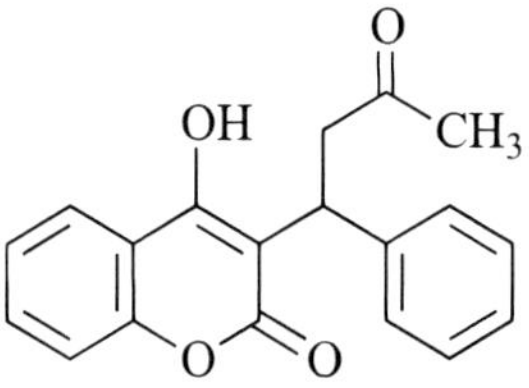   warfarin (55)

Figure 17. Phenelzine (54) and warfarin (55).

The average daily dosage of *Panax ginseng* is 1 to 2 g of dried root. *Panax ginseng* infusion may be orally administered 3 to 4 times a day over 3 to 4 weeks [72].

3.1. Cognitive Function

Oral standardized *Panax ginseng* extract 400 mg daily was effective in improving cognitive function [73].

3.2. Hypoglycemic Effects

Panax ginseng extract (100 or 200 mg) showed potent hypoglycemic effects for type 2 diabetic patients with non-insulin-dependent diabetes mellitus (NIDDM) [74].

3.3. Antiviral Effects

Standardised G115 ginseng extract (100 mg) of standardized *Panax ginseng* root extract showed higher immune response in vaccination against influenza, using total of 227 volunteers for a period of 12 weeks [30].

3.4. Erectile Dysfunction (E.D. Impotence) Effects

Standardized *Panax ginseng* (Korean red ginseng) root extract (600 mg orally three times daily) was effective for erectile dysfunction (E.D.) or impotence [75].

Panax ginseng (Korean red ginseng) root extract (900 mg orally three times daily) was effective for erectile dysfunction (E.D.) or impotence [32, 76].

Panax ginseng (Korean red ginseng) root extract (1000 mg orally three times daily) was effective for erectile dysfunction (E.D.) or impotence [35].

Panax ginseng (Korean red ginseng) root extract (1 tablet (300 mg) x 2 tablets x 3 times/day = 1800 mg orally three times daily) was effective for erectile dysfunction (E.D.) or impotence [77, 78].

3.5. Physical and Psychological Performance Capacity (Lack of Stamina)

Standardized *Panax ginseng* (Korean red ginseng) root extract (600 mg orally twice daily) was effective for improvement of physical and psychological performance capacity (lack of stamina) [79].

Daily dosage for improvement of physical and psychological performance capacity (lack of stamina) of ginseng was 2 mL extract (acute) for 40% ethanol tincture, dose unknown (12 weeks), dose unknown (90 days), 200 mg (4 weeks, 9 weeks, 12 weeks), 400 mg (20 weeks) for standardized extract, dose unknown (12 weeks) for two types of standardized extract, 200 mg (9 weeks) for standardized extract, 4% or 7% ginsenoside content, 1200 mg (3 days) for Korean white ginseng powder, 2000 mg (4 weeks) for 1.5% glycosides, 2 capsules for 30 days, 1 capsule for 30 days (60 days) for ARM229 standardized extract, 80 mg (8 weeks) for standardized extract, vitamins, minerals, 1000 mg (6 weeks) for ginseng root powder, 200 mg (6 weeks) for standardized extract plus DMAE, vitamins, minerals, and 200 mg (12 weeks) for standardized extract plus vitamins, minerals [80, 81].

For promoting utilization of free fatty acids and improving exhaustive cycling test performance in humans, daily oral administration of four capsules (total 2 g) of **STPG** complex consisting of each soybean peptides, taurine, Pueraria isoflavone, and ginseng saponin per day (500 mg/capsule) for 15 days was effective [58].

CONCLUSION

The medicinal use of *Panax ginseng* (Korean ginseng) is thought to originate from China and the Korean Peninsula. Since ancient times, *Panax ginseng* (Korean ginseng) has been prized as a therapeutic drug for wide variety of diseases although it is rare and effective in these areas. Therefore, this review is aimed to explain effective medicinal ingredients for each disease of *Panax ginseng* (Korean ginseng), and their pharmacological effects and mechanism. The diseases and preventive treatments specifically treated in this review are myocardial relaxation, antiobesity effect, inhibition of angiogenesis, hypoglycemic effects, antiviral effects, erectile dysfunction (E.D.) or impotence, improvement effects of ginseng saponins (ginsenosides), effect on exercise performance, immunomodulating activities, memory enhancing and neuroprotective effects as well as dosage and avoidance for administration. Improvement of these diseases and symptoms will be one of the top priorities to overcome in modern society. Based on the medicinal ingredients contained in *Panax ginseng* (Korean ginseng) which has been published so far, the effects and mechanisms of these medicinal ingredients are explained depending on disease prevention and treatment experiment results *etc*. In any case, since *Panax ginseng* (Korean ginseng) has various effects, it is necessary to explain in detail to researchers about the listed diseases and effects.

REFERENCES

[1] Fujitani N. Chemical and pharmacological study of Chinese medicinal ginseng. *Kyoto Medical J* 2(3), 43-83, 1905. In Japanese.
[2] https://en.oxforddictionaries.com/definition/panaquilon.
[3] Fujita M, Itokawa H, Shibata S. Chemical studies on Ginseng. I. (Studies on saponin-bearing drugs. IV.). Isolation of saponin and sapogenin from radix ginseng. *Yakugaku Zasshi* 82(12), 1634-1638, 1962.
[4] Garriques S. On panaquilon, a new vegetable substance. *Ann Chem Pharm* 90(2), 231-234, 1854.
[5] WHO. *Obesity and overweight. Fact sheet.* Updated June 2016. <http://www.who.int/mediacentre/factsheets/fs311/en/>.
[6] Friedman HS (ed.). *Encyclopedia of Mental Health.* 158, 2nd Edition, Academic Press. 2 edition, 2015.
[7] Dibaise JK, Foxx-Orenstein AE. Role of the gastroenterologist in managing obesity. *Expert Rev Gastroenterol Hepatol* 7(5), 439-451, 2013.
[8] Woodhouse R. Obesity in art: a brief overview. *Front Horm Res* 36, 271-286, 2008.
[9] Golay A, Ybarra J. Link between obesity and type 2 diabetes. *Best Pract Res Clin Endocrinol Metab* 19(4), 649-663, 2005.
[10] Mollah ML, Kim GS, Moon HK, Chung SK, Cheon YP, Kim JK, Kim KS. Antiobesity effects of wild ginseng (*Panax ginseng* C.A. Meyer) mediated by PPAR-gamma, GLUT4 and LPL in ob/ob mice. *Phytother Res* 23(2), 220-225, 2009.
[11] Ha YW, Lim SS, Ha IJ, Na YC, Seo JJ, Shin H, Son SH, Kim YS. Preparative isolation of four ginsenosides from Korean red ginseng (steam-treated *Panax ginseng* C. A. Meyer), by high-speed counter-current chromatography coupled with evaporative light scattering detection. *J Chromatogr A* 1151(1-2), 37-44. 2007.
[12] Ha DC, Lee JW, Ryu GH. Change in ginsenodides and maltol in dried raw ginseng during extrusion process. *Food Sci Biotechnol* 14(3), 363-367, 2005.

[13] Bae EA, Han MJ, Choo MK, Park SY, Kim DH. Metabolism of 20(*S*)- and 20(*R*)-ginsenoside Rg3 by human intestinal bacteria and its relation to *in vitro* biological activities. *Biol Pharm Bull* 25(1), 58-63, 2002.
[14] Wang CZ, Zhang B, Song WX, Wang A, Ni M, Luo X, Aung HH, Xie JT, Tong R, He TC, Yuan CS. Steamed American ginseng berry: ginsenoside analyses and anticancer activities. *J Agric Food Chem* 54(26), 9936-9942 2006.
[15] Wang CZ, Aung HH, Ming Ni, Wu JA, Tong R, Wicks S, He TC, Yuan CS. Red American ginseng: ginsenoside constituents and antiproliferative activities of heat-processed *Panax quinquefolius* roots. *Planta Med* 73(7), 669-674, 2007.
[16] Yoon SR, Lee GD, Park JH, Lee IS, Kwon JH. Ginsenoside composition and antiproliferative activities of explosively puffed ginseng (*Panax ginseng* C.A. Meyer). *J Food Sci* 75(4), C378-C382, 2010.
[17] Sagar SM, Yance D, Wong RK. Natural health products that inhibit angiogenesis: a potential source for investigational new agents to treat cancer-Part 1. *Curr Oncol* 13(1), 14-26, 2006.
[18] Sato K, Mochizuki M, Saiki I, Yoo YC, Samukawa K, Azuma I. Inhibition of tumor angiogenesis and metastasis by a saponin of Panax ginseng, ginsenoside-Rb2. *Biol Pharm Bull* 17(5), 635-639, 1994.
[19] Akao T, Kanaoka M, Kobashi K. Appearance of compound K, a major metabolite of ginsenoside Rb1 by intestinal bacteria, in rat plasma after oral administration--measurement of compound K by enzyme immunoassay. *Biol Pharm Bull* 21(3), 245-249, 1998.
[20] Joh EH1, Lee IA, Jung IH, Kim DH. Ginsenoside Rb1 and its metabolite compound K inhibit IRAK-1 activation--the key step of inflammation. *Biochem Pharmacol* 82(3), 278-286, 2011.
[21] Ming Y, Chen Z, Chen L, Lin D, Tong Q, Zheng Z, Song G. Ginsenoside compound K attenuates metastatic growth of hepatocellular carcinoma, which is associated with the translocation

of nuclear factor-κB p65 and reduction of matrix metalloproteinase-2/9. *Planta Med* 77(5), 428-433, 2011.

[22] Shin KO, Seo CH, Cho HH, Oh S, Hong SP, Yoo HS, Hong JT, Oh KW, Lee YM. Ginsenoside compound K inhibits angiogenesis via regulation of sphingosine kinase-1 in human umbilical vein endothelial cells. *Arch Pharm Res* 37(9), 1183-1192, 2014.

[23] Sotaniemi EA, Haapakoski E, Rautio A. Ginseng therapy in non-insulin-dependent diabetic patients. *Diabetes Care* 18(10), 1373-1375, 1995.

[24] Chan JY, Leung PC, Che CT, Fung KP. Protective effects of an herbal formulation of Radix Astragali, Radix Codonopsis and Cortex Lycii on streptozotocin-induced apoptosis in pancreatic beta-cells: an implication for its treatment of diabetes mellitus. *Phytother Res* 22(2), 190-196, 2008.

[25] Wang L, Waltenberger B, Pferschy-Wenzig EM, Blunder M, Liu X, Malainer C, Blazevic T, Schwaiger S, Rollinger JM, Heiss EH, Schuster D, Kopp B, Bauer R, Stuppner H, Dirsch VM, Atanasov AG. Natural product agonists of peroxisome proliferator-activated receptor gamma (PPARγ): a review. *Biochem Pharmacol* 92(1), 73-89, 2014.

[26] Hotamisligil GS, Shargill NS, Spiegelman BM. Adipose expression of tumor necrosis factor-alpha: direct role in obesity-linked insulin resistance. *Science* 259(5091), 87-91, 1993.

[27] Fonseca SG, Burcin M, Gromada J, Urano F. Endoplasmic reticulum stress in beta-cells and development of diabetes. *Curr Opin Pharmacol* 9(6), 763-770, 2009.

[28] Unger RH. Glucagon and the insulin: glucagon ratio in diabetes and other catabolic illnesses. *Diabetes* 20(12), 834-838, 1971.

[29] Yeo J, Kang YM, Cho SI, Jung MH. Effects of a multi-herbal extract on type 2 diabetes. *Chin Med* 2011 Mar 4;6.10.

[30] Scaglione F, Cattaneo G, Alessandria M, Cogo R. Efficacy and safety of the standardised Ginseng extract G115 for potentiating vaccination against the influenza syndrome and protection against the common cold [corrected]. *Drugs Exp Clin Res* 22(2), 65-72, 1996.

[31] Choi HK, Seong DH, Rha KH. Clinical efficacy of Korean red ginseng for erectile dysfunction. *Int J Impot Res* 7(3), 181-186, 1995.

[32] Hong B, Ji YH, Hong JH, Nam KY, Ahn TY. A double-blind crossover study evaluating the efficacy of Korean red ginseng in patients with erectile dysfunction: a preliminary report. *J Urol* 168(5), 2070-2073, 2002.

[33] Tachikawa E, Kudo K, Harada K, Kashimoto T, Miyate Y, Kakizaki A, Takahashi E. Effects of ginseng saponins on responses induced by various receptor stimuli. *Eur J Pharmacol* 369(1), 23-32, 1999.

[34] O'Hara M, Kiefer D, Farrell K, Kemper K. A review of 12 commonly used medicinal herbs. *Arch Fam Med* 7(6), 523-536, 1998.

[35] de Andrade E, de Mesquita AA, Claro Jde A, de Andrade PM, Ortiz V, Paranhos M, Srougi M. Study of the efficacy of Korean Red Ginseng in the treatment of erectile dysfunction. *Asian J Androl* 9(2), 241-244, 2007.

[36] Oliynyk S, Oh S. The pharmacology of actoprotectors: practical application for improvement of mental and physical performance. *Biomol Ther* (Seoul) 20(5), 446-456, 2012.

[37] Wagner JC. Abuse of drugs used to enhance athletic performance. *Am J Hosp Pharm* 46(10), 2059-2067, 1989.

[38] Liu CX, Xiao PG. Recent advances on ginseng research in China. *J Ethnopharmacol* 36(1), 27-38, 1992.

[39] Brekhman II, Dardymov IV. New substances of plant origin which increase nonspecific resistance. *Annu Rev Pharmacol* 9, 419-430, 1969.

[40] Saito H, Yoshida Y, Takagi K. Effect of *Panax ginseng* root on exhaustive exercise in mice. *Jpn J Pharmacol* 24(1), 119-127, 1974.

[41] Bahrke MS, Morgan WP. Evaluation of the ergogenic properties of ginseng. *Sports Med* 18(4), 229-248, 1994.

[42] Carr CJ, Jokl E. (eds.) In: *Enhancers of Performance and Endurance: A Symposium*. pp. 138-192, 1986. Lawrence Erlbaum Associates (Publishers), NJ, USA.

[43] Wang LC, Lee TF. Effect of ginseng saponins on exercise performance in non-trained rats. *Planta Med* 64(2), 130-133, 1998.

[44] Aoyama T, Fukui K, Takamatsu K, Hashimoto Y, Yamamoto T. Soy protein isolate and its hydrolysate reduce body fat of dietary obese rats and genetically obese mice (yellow KK). *Nutrition* 16(5), 349-354, 2000.
[45] Claessens M, Calame W, Siemensma AD, Saris WH, van Baak MA. The thermogenic and metabolic effects of protein hydrolysate with or without a carbohydrate load in healthy male subjects. *Metabolism* 56(8), 1051-1059, 2007.
[46] Drăgan I, Stroescu V, Stoian I, Georgescu E, Baloescu R. Studies regarding the efficiency of Supro isolated soy protein in Olympic athletes. *Rev Roum Physiol* (1990) 29(3-4), 63-70, 1992.
[47] Huxtable RJ. Physiological actions of taurine. *Physiol Rev* 72(1), 101-163, 1992.
[48] Nieminen ML, Tuomisto L, Solatunturi E, Eriksson L, Paasonen MK. Taurine in the osmoregulation of the Brattleboro rat. *Life Sci* 42(21), 2137-2143, 1988.
[49] Manabe S, Kurroda I, Okada K, Morishima M, Okamoto M, Harada N, Takahashi A, Sakai K, Nakaya Y. Decreased blood levels of lactic acid and urinary excretion of 3-methylhistidine after exercise by chronic taurine treatment in rats. *J Nutr Sci Vitaminol (Tokyo)* 49(6), 375-380, 2003.
[50] Shi RL, Zhang JJ. Protective effect of puerarin on vascular endothelial cell apoptosis induced by chemical hypoxia in vitro. *Yao Xue Xue Bao* 38(2), 103-107, 2003. in Chinese.
[51] Wang LC, Lee TF. Effect of ginseng saponins on exercise performance in non-trained rats. *Planta Med* 64(2), 130-133, 1998.
[52] Fitts RH, Holloszy JO. Lactate and contractile force in frog muscle during development of fatigue and recovery. *Am J Physiol* 231(2), 430-433, 1976.
[53] Allen D, Westerblad H. Physiology. Lactic acid--the latest performance-enhancing drug. *Science* 305(5687), 1112-1113, 2004.
[54] Kristensen M, Albertsen J, Rentsch M, Juel C. Lactate and force production in skeletal muscle. *J Physiol* 562(Pt 2), 521-526, 2005.

[55] Cairns SP. Lactic acid and exercise performance: culprit or friend? *Sports Med* 36(4), 279-291, 2006.
[56] McGarry JD, Brown NF. The mitochondrial carnitine palmitoyltransferase system. From concept to molecular analysis. *Eur J Biochem* 244(1), 1-14, 1997.
[57] Wang LC, Lee TF. Effect of ginseng saponins on exercise performance in non-trained rats. *Planta Med* 64(2), 130-133, 1998.
[58] Yeh TS, Chan KH, Hsu MC, Liu JF. Supplementation with soybean peptides, taurine, Pueraria isoflavone, and ginseng saponin complex improves endurance exercise capacity in humans. *J Med Food* 14(3), 219-225, 2011.
[59] Bu H, Nie L, Wang D, Yuan S, Li S, Guo Z, Xu X, Wang G, Li X. Rapid determination of *Panax ginseng* by near-infrared spectroscopy. *Analytical Methods* 5(23), 6715-6721, 2013.
[60] Guo L, Liu J, Hu Y, Wang D, Li Z, Zhang J, Qin T, Liu X, Liu C, Zhao X, Fan YP, Han G, Nguyen TL. Astragalus polysaccharide and sulfated epimedium polysaccharide synergistically resist the immunosuppression. *Carbohydr Polym* 90(2), 1055-1060, 2012.
[61] Fan Y, Lu Y, Wang D, Liu J, Song X, Zhang W, Zhao X, Nguyen TL, Hu Y. Effect of epimedium polysaccharide-propolis flavone immunopotentiator on immunosuppression induced by cyclophosphamide in chickens. *Cell Immunol* 281(1), 37-43, 2013.
[62] Liu LN, Guo ZW, Zhang Y, Qin H, Han Y. Polysaccharide extracted from *Rheum tanguticum* prevents irradiation-induced immune damage in mice. *Asian Pac J Cancer Prev* 13(4), 1401-1405, 2012.
[63] Kaplan SS, Lancaster JR Jr, Basford RE, Simmons RL. Effect of nitric oxide on staphylococcal killing and interactive effect with superoxide. *Infect Immun* 64(1), 69-76, 1996.
[64] Foster S. Echinacea. The cold and flu remedy. *Alternative & Complementary Therapies* 1(4), 254–257, 1995.
[65] Lim TS, Na K, Choi EM, Chung JY, Hwang JK. Immunomodulating activities of polysaccharides isolated from *Panax ginseng*. *J Med Food* 7(1), 1-6, 2004.

[66] Petkov VD, Mosharrof AH. Effects of standardized ginseng extract on learning, memory and physical capabilities. *Am J Chin Med* 15(1-2), 19-29, 1987.
[67] Zhong YM, Nishijo H, Uwano T, Tamura R, Kawanishi K, Ono T. Red ginseng ameliorated place navigation deficits in young rats with hippocampal lesions and aged rats. *Physiol Behav* 69(4-5), 511-525, 2000.
[68] Kennedy DO, Scholey AB. Ginseng: potential for the enhancement of cognitive performance and mood. *Pharmacol Biochem Behav* 75(3), 687-700, 2003.
[69] Bao HY, Zhang J, Yeo SJ, Myung CS, Kim HM, Kim JM, Park JH, Cho J, Kang JS. Memory enhancing and neuroprotective effects of selected ginsenosides. *Arch Pharm Res* 28(3), 335-342, 2005.
[70] Kiefer D, Pantuso T. *Panax ginseng. Am Fam Physician* 68(8), 1539-1542, 2003.
[71] Mahady GB, Gyllenhall C, Fong HH, Farnsworth NR. Ginsengs: a review of safety and efficacy. *Nutr Clin Care* 3(2), 90-101, 2000.
[72] Joerg; Economics, Medical; Reference, Pdr Physicians Desk Gruenwald (Author). *PDR for Herbal Medicines*, Third Edition. 2004, p.383. Thomson Healthcare (Publisher), Montvale, NJ, USA.
[73] Sørensen H, Sonne J. A double-masked study of the effects of ginseng on cognitive functions. *Curr Ther Res* 57(12), 959-968, 1996.
[74] Sotaniemi EA, Haapakoski E, Rautio A. Ginseng therapy in non-insulin-dependent diabetic patients. *Diabetes Care* 18(10), 1373-1375, 1995.
[75] Choi HK, Seong DH, Rha KH. Clinical efficacy of Korean red ginseng for erectile dysfunction. *Int J Impot Res* 7(3), 181-186, 1995.
[76] Kim SW, Paick JS. Clinical efficacy of Korean red ginseng on vasculogenic impotent patients. *Korean J Androl* 17(1), 23–28, 1999.
[77] Choi HK, Choi YJ, Kim JH. Penile blood change after oral medication of Korean red ginseng in erectile dysfunction patients. *J Ginseng Res* 27(4), 165-170, 2003.

[78] Choi HK, Choi YD, Adaikan PG, Jiang Y. Effectiveness of Korean red ginseng in erectile dysfunction: multi-national approach. *J Ginseng Res* 23(4), 247–256, 1999.

[79] Forgo I, Schimert G. On the question of the duration of the standardized ginseng extract G 115 healthy performers. *Notabene Medici* 9, 636-649, 1985.

[80] Bucci LR. Selected herbals and human exercise performance. *Am J Clin Nutr* 72(2 Suppl), 624S-636S, 2000.

[81] Oliynyk S, Seikwan O. Actoprotective effect of ginseng: improving mental and physical performance. *J Ginseng Res* 37(2), 144-166, 2013.

In: Occurrences, Structure, Biosynthesis ... ISBN: 978-1-53614-141-2
Editor: Noboru Motohashi © 2018 Nova Science Publishers, Inc.

Chapter 2

COFFEE'S PHYTOCHEMICALS: FROM BIOSYNTHESIS TO HEALTH BENEFITS

Lourdes Valadez-Carmona[1,*],
Carla Patricia Plazola-Jacinto[1,†],
Marcela Hernández-Ortega[2,‡]
and D. Nayelli Villalón-López[3,§]

[1]Departamento de Ingeniería Bioquímica, Escuela Nacional de Ciencias Biológicas, Instituto Politécnico Nacional, México
[2]Facultad de Ciencias de la Salud,
Universidad Anáhuac México Norte, México
[3]Departamento de Química Orgánica, Escuela Nacional de Ciencias Biológicas, Instituto Politécnico Nacional, México

ABSTRACT

Coffee is a popular beverage produced from the roasted beans of a great variety of coffee crops, being the most economically important

---

[*] lvc24@hotmail.com.
[†] patricia.plazola@gmail.com.
[‡] marcelahdz17@yahoo.com.mx.
[§] Corresponding Author: lorienaule@gmail.com.

species *Coffea canephora* and *Coffea arabica*. The brews of these crops are worldwide consumed due to their good organoleptic qualities, which are related to the presence of different molecules such as caffeine, polyphenols, melanoidins and carbohydrates. Besides its sensory likeable characteristics, coffee brew possesses important beneficial health activities such as antiproliferative effect against some human cancer cell lines, therapeutic potential against some neurodegenerative illness and antioxidant capacities, among others. Because of the different beneficial effects related to the coffee, many analytical techniques have been studied in order to identify and quantify the phytochemicals to safeguard the well-being of the consumer.

Keywords: coffee brew, phytochemicals, diterpenes, caffeine, chlorogenic acid

CHEMICAL CONSTITUENTS

- cafestol (CAF. **1**)
- kahweol (KAH. **2**)
- chlorogenic acid (CGA. 3-caffeoylquinic acid. **3**)
- caffeine (1,3,7-trimethylxanthine. **4**)
- trigonelline (**5**)
- shikimic acid (**6**)
- *p*-coumaric acid (**7**)
- caffeic acid (**8**)
- ferulic acid (**9**)
- caffeoylquinic acid (**10**)
- (-)-quinic acid (**11**) add structure
- *p*-feruloylquinic acid (**12**)
- dicaffeoylquinic acid (**13**)
- *p*-coumaroylquinic acid (**14**)
- caffeoyl-D-glucose (**15**)
- caffeoyl-CoA (**16**)
- 5-*O*-caffeoylquinic acid (neochlorogenic acid. 5-CQA. **17**)

- xanthosine (**18**)
- 7-methylxanthosine (**19**)
- ent-kaurene (**20**)
- hydroxycinnamic acid (**21**)
- glucose (**22**)
- homocysteine (**23**)
- metformin (**24**)

ABBREVIATIONS

7-MXS	7-methylxanthosine synthase
AFB1	aflatoxin B1
ALT	alanine aminotransferase
AST	aspartate aminotransferase
BMR	basal metabolic rate
C3H	*p*-coumarate 3'-hydroxylase
CE	catechin equivalent
CGAs	chlorogenic acids
CNS	central nervous system
CPT1	carnitine palmitoyltransferase I
CS	caffeine synthase
CYP	cytochrome P450
ELAM-1	E-selectin endothelial leukocyte adhesion molecule-1
ent-CPP	ent-copalyl diphosphate
FAS	fatty acid synthase
G6Pase	glucose-6-phosphatase
GAE	gallic acid equivalent
GGPP	geranylgeranyl diphosphate
GGT	gamma-glutamyl transferase
GIP	insulinotropic polypeptide
GK	glucokinase

GLP-1	glucagon-like peptide-1
Gpx	glutathione peroxidase
GR	glutathione reductase
GST	glutation S-transferase
HDL-c	high-density lipoprotein cholesterol
HPLC	high performance liquid chromatography
HQT	hydroxycinnamoyl-coenzyme A quinate transferases
ICAM-1	intercellular adhesion molecule-1
IPP	isopentenyl diphosphate
KS	kaurene synthase
LDL-c	low-density lipoprotein cholesterol
MAE	microwave-assisted extraction
MMPs	endo-peptidases; matrix metalloproteinases
MMP-1	matrix metalloproteinase 1
MMP-2	matrix metalloproteinase 2
MMP-9	matrix metalloproteinase 9
MTBE	*tert*-butyl ether
NDMA	N-nitrosodimethylamine
Nrf2	nuclear respiratory factor 2
ODS	reverse-phased column C18
PhIP	2-amino-1-methyl-6-phenylimidazopyridine
PLE	pressurized liquid extraction
Rf	retention factor
SAM	S-adenosylmethionine
SC-CO2	supercritical carbon dioxide extraction
SFE	supercritical fluid extraction
TLC	thin layer chromatography
TS	theobromine synthase
UV	ultraviolet
VCAM-1	vascular cell adhesion molecule-1

1. INTRODUCTION

Coffee is the infusion obtained from ground roasted or un-roasted coffee beans; that has taken an important place in human society, being one of the most consumed beverages worldwide. In fact, coffee consumption originated probably in northeast Africa and spread out in the 15th century through the Middle East and Europe. Today, coffee consumption has become a regular part of daily life [1, 2]. The consumption of coffee were 42,604 in the European Union; 25,336 in the USA and 2,354 in Mexico of bags of 60 kg [3].

Coffee aroma and flavor are the result of a complex chemical mixture of more than 2,000 chemical compounds, including carbohydrates, lipids, nitrogenous compounds, vitamins, minerals, alkaloids and phenolic compounds that give these organoleptic characteristics to the beverage, making the coffee the main source of caffeine in many populations [1, 4, 5, 6].

The relationship between coffee and health has been reported in many studies for over the last 40 years, among these studies, controversies exist about the positive or negative effects on consumers' health. Epidemiological data support that habitual coffee consumption is associated with several health benefits such as the reduction of type 2 diabetes development, reduction of liver damage in people at high risk for liver disease, reduction of ischemic heart disease risk, risk reduction of strokes, risk reduction of several types of cancer and depression [2, 7, 8, 9, 10]. Furthermore, coffee consumption has showed stimulatory effects on the central nervous system (CNS) with a positive long-term memory effect and reducing the risk of Parkinson's disease (PD) [1, 2, 8, 10]. Coffee health benefits have been attributed to its bioactive compounds such as chlorogenics acids (caffeoylquinic acid, feruloylquinic acid, *p*-coumaroylquinic acid), flavonoids (catechins, anthocyanins), tocopherols, hydroxycinnamic acids (ferulic acid, caffeic acid and *n*-cumaric acid), methylxanthines (caffeine, theobromine, theophylline), diterpene alcohols (cafestol, kahweol), vitamin B3 (precursor of trigonelline), niacin, and melanoidins [6, 11, 12].

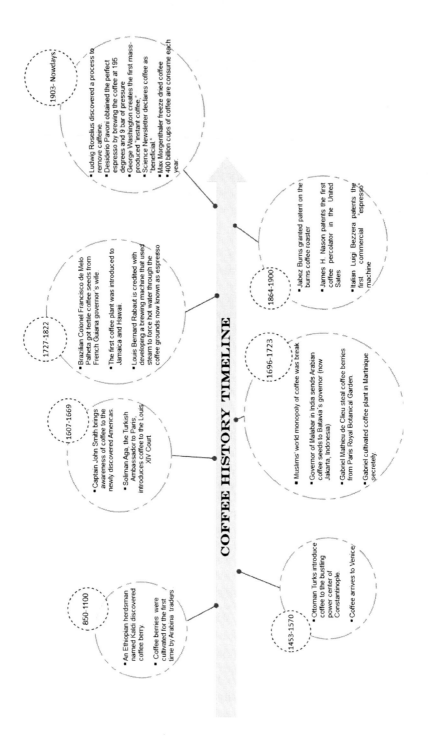

Figure 1. Coffee history timeline.

2. HISTORY AND TIMELINE

The word 'coffee' name is derived from the Arabic word *quahweh* and from Latin, *coffea* for its botanical genus [8]. The history of coffee begins in Kefa in Northern Africa, today known as Ethiopia circa 850 AD (Figure 1). The coffee discovery coffee was accidental, when Kaldi herded his goats and noticed that they were acting in a frantic way after eating red berries from a small, broad-leaved shrub [13, 14, 15]. Currently we know that these berries are coffee cherries in which the green coffee beans are cover by a red fleshy berry. At the beginning, coffee was not consumed as beverage, but as high-energy snack mixed with fat for long journeys [13]. Coffee seems to have been brought from Ethiopia to the Port of Mocha through the trade route between Arabia and East Africa.

3. TAXONOMY, PRODUCTION AND COFFEE BEANS PROCESSING

Coffee tree (*Rubiaceae* family, *Coffea* genus) is an economical valuable agricultural crop due to the seeds contained in its berries [6, 16]. The coffee fruit consists of a smooth, tough outer skin or pericarp, being usually green in the unripe fruits, but turning red-violet or deep red when they ripe. This pericarp covers a soft yellowish, fibrous and sweet pulp; between this pulp and the yellowish endocarp, there is a translucent, colorless, thin, viscous and highly hydrated layer of mucilage. The silverskin covers each hemisphere of the coffee bean (Figure 2) [17].

Coffee is cultivated in tropical and subtropical countries located between the latitudes of Cancer (23° 26′ north of the equator) and Capricorn (23° 26′ south of the equator) [18]. The commercial production exploits two major species, arabica (*Coffea arabica* L.) which represents approximately 70% of the world market and is produced predominantly in the American Continent, and robusta (*Coffea canephora*) that account for approximately 30% of the production and it is mainly grown in Asia and

Africa. Arabicas are native from the Ethiopian highlands, whereas robustas originated at lower altitudes across the Ivory Coast, Congo and Uganda [13, 19, 20]. In 2006, the major coffee producers, according to statistics of *the International Coffee Organization, were* Brazil, Vietnam, Colombia, Indonesia, Ethiopia, Honduras, India, Peru, Uganda, Guatemala and Mexico [3].

To obtain the seeds from the harvested fruit, there are two methods, the wet and the dry. The dry process is used principally for robustas in which the freshly cherries are sun dried for 2-3 weeks, then, the dried husk is removed mechanically, whereas in the wet process, the pulp is removed prior to drying by soaking into water and fermenting the cherries. After the drying process, the beans that are able to be commercialized can be roasted at different combinations of temperatures and times, the most used combinations are ~230°C for few minutes, or at ~180°C for up to 20 minutes to obtain a dark-brown/black color. Under these conditions, chemical reactions such as exothermic pyrolysis and myriad reactions occur, these reactions are responsible for the development of the characteristic aroma and taste of coffee [13]. After roasting, most of the robustas are blended, they are preferred for instant coffee production, since they are generally considered inferior to arabicas.

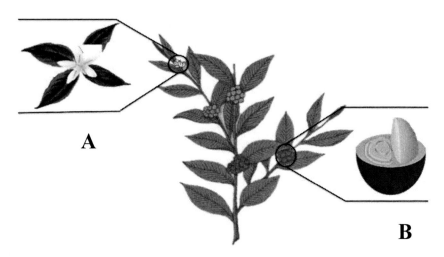

Figure 2. Parts of the coffee plant. A: coffee flower; B: coffee bean.

4. PHYTOCHEMICALS AND BIOSYNTHESIS

Table 1. Major phytochemicals content present in coffee beans

Compound	Structure	Content (%)
chlorogenic acids (CGAs) - chlorogenic acid (CGA. 3)		6 – 10 [13]
caffeine (4)		1.45 – 2.38 [20]
trigonelline (5)		1.01 -1.39 [20]
diterpenes - cafestol (CAF. 1)		0.2 - 1.9 [22]
- kahweol (KAH. 2)		

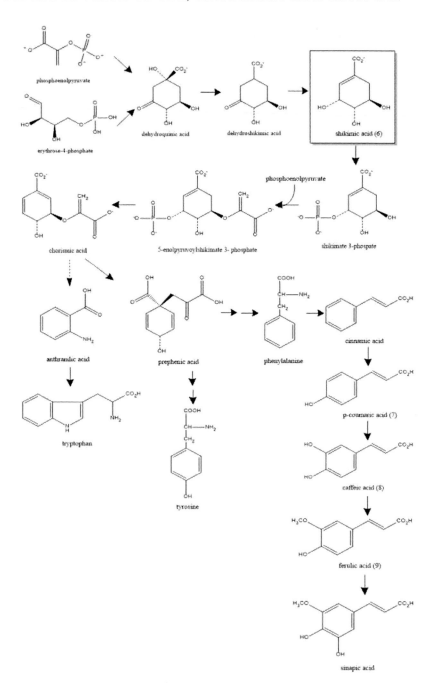

Figure 3. Phenolic compounds biosynthesis via the shikimic acid pathway. Shikimic acid (**6**), p-coumaric acid (**7**), caffeic acid (**8**), and ferulic acid (**9**).

Coffee has been recognized as a functional beverage due to the high amounts of bioactive compounds such as phenolic compounds, caffeine, melanoidins, trigonelline, cafestol (CAF. 1) and kahweol (KAH. 2) among others (Table 1); being an important source of antioxidants in the human diet [21]. The most important bioactive chemical constituents of coffee involve phenolic compounds and some of their derivatives such as chlorogenic acids (CGA) including chlorogenic acid (CGA. 3) (Table 1), some alkaloids, diterpenoids, alcohols, carbohydrates, lipids and volatile heterocyclic compounds.

The biosynthesis of phenolic compounds is regulated by phenylalanine ammonia lyase (PAL), an enzyme which its activity varies across the growth cycle of the plant. Most of the phenylalanine can be used in the formation of phenolic compounds when there is no fertile cycle. Shikimic acid (6) (Figure 3) is the main precursor to synthesize phenolic compounds *via* glucose cyclization (Figure 3) [23].

4.1. Phenolic Compounds-Chlorogenic Acids

Simple phenolic compounds are synthesized from phenylalanine catalyzed by phenylalanine ammonia lyase (PAL) generating cinnamic acid, in presence of hydroxylases the formation of *p*-coumaric acid (7) takes place. The major phenolic compound in coffee is the caffeoylquinic acid (10) (Figure 5), which is part of a family of esters between certain *trans*-cinnamic acids, such as caffeic acid (8) and ferulic acid (9), with (-)-quinic acid (11) (Figure 5), and is often referred to as chlorogenic acid (CGA. 3) [4, 16, 24, 25]. However, the term 'chlorogenic acids' (CGAs) stand for the whole set of hydroxycinnamic esters with (-)-quinic acid (11), including caffeoylquinic acid (10) (Figure 5), *p*-feruloylquinic acid (12) (Figure 5), dicaffeoylquinic acid (13) (Figure 5) and *p*-coumaroylquinic acid (14) (Figure 4) [25].

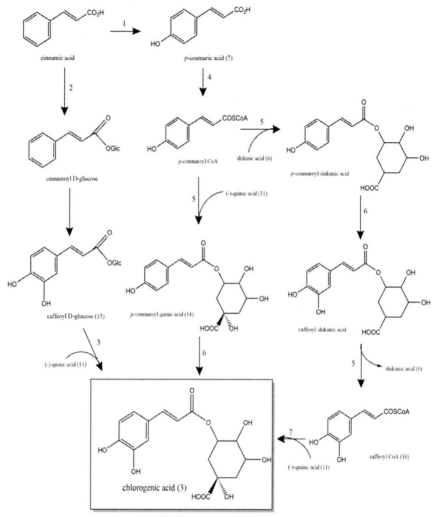

Enzimas
1) C4H: Cinnamic acid 4-hydroxylase
2) UGCT: UDP Glucose:cinnamate glucosyl transferase
3) HCGQT: Hydroxyxinnamoyl D-glucose:quinate hydroxycinnamoyl transferase
4) 4CL: 4-coumarate CoA ligase
5) HCT: Hydroxycinnamoyl CoA
6) C3H: p-coumarate 3'-hydroxylase
7) HQT: Hydroxycinnamoyl CoA quinate hydroxycinnamoyl transferase

Figure 4. Biosynthesis of chlorogenic acid (CGA. 3-caffeoylquinic acid. **3**).

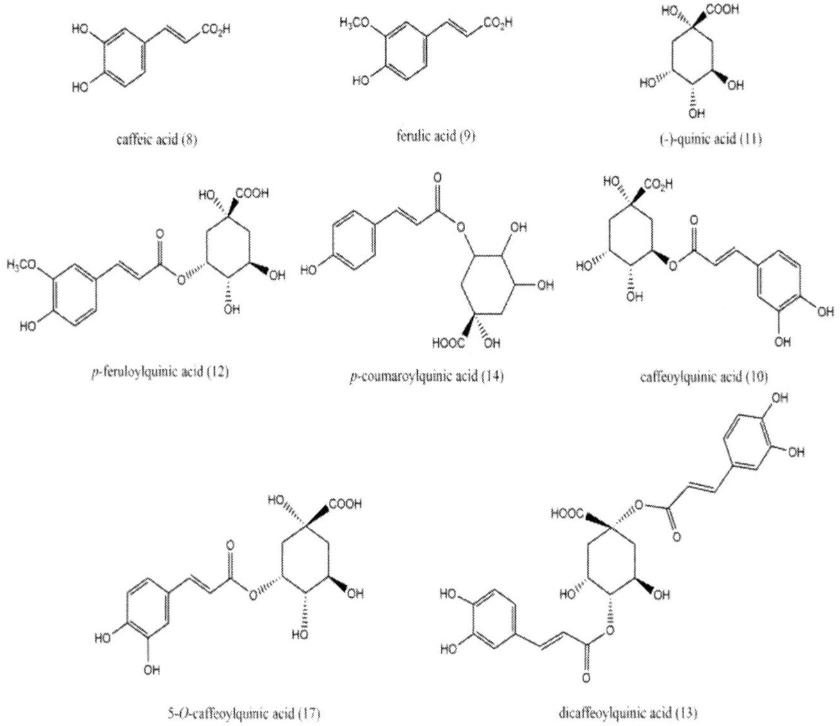

Figure 5. Caffeoylquinic acid (**10**), (-)-quinic acid (**11**), p-feruloylquinic acid (**12**), dicaffeoylquinic acid (**13**), and 5-O-caffeoylquinic acid (neochlorogenic acid. **17**).

Chlorogenic acids (CGAs) plant metabolism importance lies in the fact that they are the precursors of G- and S type lignin biosynthesis and play a prominent role in tissue senescence. The complete pathway for the synthesis of chlorogenic acid is still debated; three distinct pathways have been considered (Figure 4): i) caffeoyl-D-glucose (**15**) is used as the activated intermediate, ii) synthesis from caffeoyl-CoA (**16**) and quinic acid (**11**) by hydroxycinnamoyl-coenzyme A quinate transferases (HQT), and iii) hydroxylation of *p*-coumaroylquinic acid (**14**) by *p*-coumarate 3'-hydroxylase (C3H). The most common individual chlorogenic acids are 5-O-caffeoylquinic acid (neochlorogenic acid. **17**) and 3-caffeoylquinic acid (chlorogenic acid 3) (Figure 5). The content of chlorogenic acids (CGAs) in a 200 mL coffee cup has been reported to range from 70-350 mg and caffeine acid range from 35-175 mg (8) [4, 26].

4.2. Alkaloids

Caffeine (1,3,7-trimethylxanthine. 4) is the most important methylxanthine present in coffee, which is perceived as a stimulant of the central nervous system (CNS) and bronchodilator. Caffeine (4) consumption increases awareness and wakefulness, improves clear thinking and attenuates fatigue [27]. The caffeine (4) content in coffee beans ranges from 10-12 mg/g in *Arabica* and from 19-21 mg/g in *Robusta*, while caffeine concentration in coffee beverages is quite variable; a cup of coffee often provides 100 mg of caffeine, but recent analysis of 14 different kinds of coffees discover that the amount of caffeine (4) ranges from 72-130 mg, depending on the kind [2, 4, 26]. Caffeine (4) has beneficial or toxic effects related to its structure; moreover, it has been shown to have lipophilic characteristics, which help its diffusion through cell membranes and cross the blood-brain barrier (BBB).

The purine nucleotides form methylxanthines in plants, being the xanthosine (18) the initial substrate for their biosynthesis, which may be supplied by different pathways, including purine biosynthesis (*de novo* route), degradation of adenine nucleotide (AMP), the *S*-adenosylmethionine (SAM) cycle and guanine nucleotide (GMP) cycle. The main pathway of caffeine (4) production is the methylation of xanthosine (18), where *S*-adenosylmethionine (SAM) serves as the methyl donor (Figure 6).

The enzymes proposed to be involved in the consecutive reactions are (19) (Figure 6) 7-methylxanthosine synthase (7-MXS), *N*-methyl nucleosidase, theobromine synthase (TS) and caffeine synthase (CS). The rate of caffeine biosynthesis is primarily regulated by the induction or repression of *N*-methyltransferases, with special focus on 7-methylxanthosine synthase (7-MXS). Therefore, the rate-limiting step in the caffeine biosynthetic pathway is the initial conversion of xanthosine (18) to 7-methylxanthosine (19), catalysed by the 7-methylxanthosine synthase (Figure 6) [27, 28].

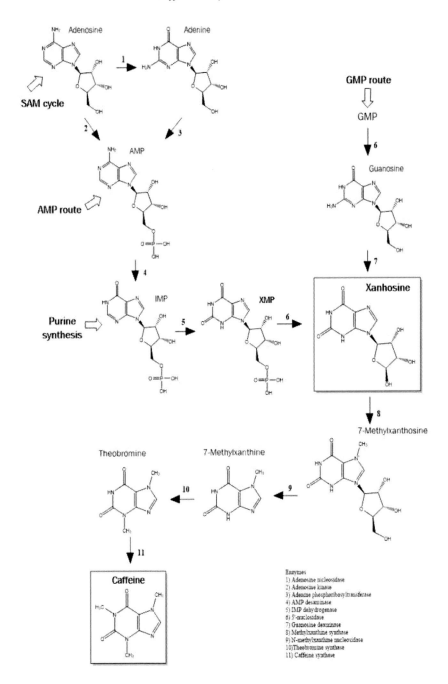

Figure 6. Biosynthesis of caffeine, (1,3,7-trimethylxanthine. 4) the principal alkaloid present in coffee.

4.3. Diterpenes

Cafestol (CAF. 1), 16-*O*-methyl-cafestol and kahweol (KAH. 2) are pentacyclic diterpenes and the major components of the unsaponifiable fraction in green and roasted coffee beans, with a characteristic furan group; both are unique chemicals to coffee beans and brews and rarely present in their free form, they commonly exist in larger quantities, esterified with fatty acids (mainly palmitic acid and linoleic acid) [29, 30, 31, 32].

Arabicas contains about 0.3-0.7% w/w of cafestol (CAF. 1) and kahweol (KAH. 2), whereas robustas has less concentration about 0.1-0.3% of cafestol (CAF. 1) and traces of kahweol (KAH. 2) (<0.01%); however, the amount of these compounds depends on the type of coffee, as well as on the preparation method [29, 33].

Since cafestol (CAF. 1) and kahweol (KAH. 2) are probable ent-kaurene derivatives, from their biosynthesis pathway they are responsible for the isopentenyl diphosphate (IPP) precursor. IPP is used as a substrate to generate geranylgeranyl diphosphate (GGPP); thereafter, a copalyl diphosphate synthase catalyzes the cyclization of GGPP to ent-copalyl diphosphate (ent-CPP), which is converted to ent-kaurene (20) (Figure 7) by kaurene synthase (KS) in several plants [30].

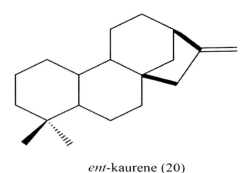

ent-kaurene (20)

Figure 7. Structure of *ent*-kaurene (20).

5. ROASTING EFFECT ON COFFEE'S PHYTOCHEMICALS

Green coffee beans contain large quantities of phenolic compounds, mainly chlorogenic acids (CGAs), and water-soluble phenols, which are formed by the esterification of (-)-quinic acid (11) with one or more cinnamic acids [34]; there are some other molecules that are not present in green coffee (Figure 8), but are formed during the roasting process due to several reactions such as pyrolysis, Strecker degradation, Maillard reaction and caramelization, by which the coffee beans change their sensorial characteristics (odor, flavor and aroma) [19].

During the roasting process, melanoidins are formed. Melanoidins are high molecular weight, nitrogenous brown polymers, which provide the color of coffee and possess antioxidant activity *in vitro*. Whilst melanoidins are formed, the roasting process also results in the loss of some of the chlorogenic acids (CGAs), that are either degraded or partially incorporated in the melanoidins [16, 29, 35]. Other compounds that decreased with the roasting process are diterpenes, but the production of diterpenes depends on the temperature during this process [36].

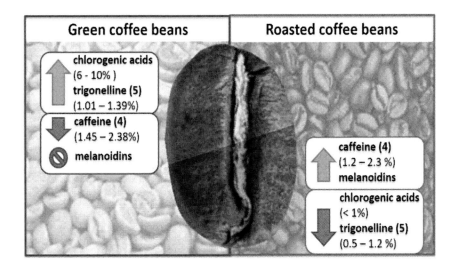

Figure 8. Phytochemical compounds affected by roasted process.

6. EXTRACTION METHODS

Different extraction methods and techniques for coffee chemicals identification have been used (Table 2). Soxhlet extraction, reflux, cold pressing, or maceration by organic solvents have been conducted as extraction processes to obtain phytochemical compounds from different sources [37]. Though, these conventional methods have limitations concerning the long extraction times, low yield and low quality of the extracts, losses of volatile compounds, degradation of the bio-compounds due to the heat and high solvent consumption [37, 38]. Therefore, it is necessary the use of an extraction technique able to overcome the mentioned limitations, and that is environmentally friendly, avoiding or minimizing the use of organic solvents, such as supercritical fluid extraction (SFE), microwave-assisted extraction (MAE), pressurized liquid extraction (PLE), ultrasounds among others [32, 36].

Araújo and Sandi [36] compared Soxhlet extraction and supercritical carbon dioxide (SC-CO_2) extraction to extract the coffee oil fraction from both green and roasted coffee beans. Soxhlet diterpenes extraction was higher (860.1 and 726.3 mg/100 g of green and roasted coffee, respectively) than SC-CO_2 extraction having 52% and 71.2% reduction of diterpene content in green and roasted coffee beans, respectively, although the extraction yield and diterpene content is lower in SC-CO_2 extraction, it might be a safer option since it is obtained without the use of organic solvents.

Belandria, Aparecida de Oliveira [39] studied the use of pressurized liquid (PLE) extraction and its optimal extraction conditions, studying the solvent, the flow rate and the temperature to extract diterpenes from green coffee beans. The optimal extraction conditions found by this study were: methanol at a flow rate of 3 mL/min at 100°C, the cafestol (CAF. 1) and kahweol (KAH. 2) extraction yields were 155% and 172%, respectively.

Table 2. Summary of some extraction methods applied to extract different phytochemical of coffee beans

Sample	Extracted compounds	Conditions tested	Results	Reference
Green and roasted coffee beans	Diterpenes	*Soxhlet* solvent: hexane (10 mL/g) 16 h *Supercritical carbon dioxide (SC-CO₂)* temperature: 60 – 90°C pressure: 235 – 380 bar	green coffee *Soxhlet*: 860.1 mg/100 g *(SC-CO₂)* 453.3 mg/100 g at 70°C and 253 bar roasted coffee *Sohxlet*: 726.3 mg/100 g *(SC-CO₂)* 209.1 mg/100 g at 70°C and 371 bar	[36]
Arabica green coffee beans		*Soxhlet* proportion extraction 1:10 for 6 h	Soxhlet total iterpenes: 712.3 mg/100 g cafestol (CAF. 1): 267.0 mg/100 g kahweol (KAH. 2): 445.3 mg/100 g	[39]
Arabica green coffee beans	Diterpenes	*Pressurized-fluid extraction (PLE) system.* solvents: water, ethanol and methanol flow rate: 0.5, 1.0, 2.0 and 3.0 mL/min extraction temperature: 60, 80 and 100°C extraction time: 165 min	PLE best conditions methanol at 100°C at a flow rate of 3.0 mL/min. total diterpenes: 1151.1 mg/100 g cafestol (CAF. 1): 461.4 mg/100 g kahweol (KAH. 2): 689.7 mg/100 g	

Table 2. (Continued)

Sample	Extracted compounds	Conditions tested	Results	Reference
Green arabica coffee beans from Brazil		*Soxhlet* petroleum ether in a ratio 4:1 with the sample, heated to reflux for 4 h. *Microwave-assisted extraction (MAE)* sample amount: 2 g solvent volume: 8 mL temperature: 30, 37.5 and 45°C extraction time: 2, 6 and 10 min	Oil Yield Soxhlet: 74.9 - 72.6 g/kg MAE best conditions (45°C/10 min) 58.6 - 76.1 g/kg Diterpenes Soxhlet: 7.26 - 9.77 g/kg MAE: 4.53 - 8.91 g/kg	[32]
Green coffee beans *(C. arabica* L. cv. Medellin) at different roasting degrees	Total phenolic compounds (TPC) and total flavonoids compounds (TFC)	*Roasting process:* 240°C time of roasting degrees: light: 9 min medium: 11 min dark: 13 min very dark: 16 min *Extraction conditions:* 50 g of each sample were dissolved in 1 L of distilled water and stirred for 5 min at 80°C	*(TPC)* unroasted: 36.93 mg GAE^1/g light-roasted: 55.00 mg GAE^1/g medium-roasted: 41.49 mg GAE^1/g dark-roasted: 35.95 mg GAE^1/g very dark-roasted: 26.80 mg GAE^1/g *(TFC)* unroasted: 20.41 mg CE^2/g light-roasted: 24.79 mg CE^2/g medium-roasted: 15.40 mg CE^2/g dark-roasted: 9.30 mg CE^2/g very dark-roasted: 4.88 mg CE^2/g	[40]

Sample	Extracted compounds	Conditions tested	Results	Reference
Mexican green coffee beans (*Coffea arabica*)	Chlorogenic acids	*Sample*: lyophilized green coffee powder *Extraction methods*: 100 g of sample with 500 mL of solvent were magnetically stirred at different conditions in each method, after they were filtered through celite	Extraction and isolation yields of chlorogenic acids *Isolation method 1*: water: 5.62% methanol: 5.87% isopropanol: 5.87%	[41]
Mexican green coffee beans (*Coffea arabica*)	Chlorogenic acids	Water: stirred for 30 min at 80°C. Aqueous methanol (70%): stirred for 24 h at room temperature. Methanol was evaporated in a rotary evaporator. Aqueous isopropanol 60%: stirred for 48 h at room temperature, in the dark. Isopropanol was evaporated in a rotary evaporator. *Isolation methods* 1) Each extract was added with ammonium sulphate until reach a concentration of 20 g/L. After this, 300 mL phosphoric acid (4%) was added. The aqueous phase was extracted 4 times with 300 mL of ethyl acetate. This phase was dried with 10 g of anhydrous sodium sulphate, filtered and evaporated in a rotary evaporator at 40°C and 120 rpm.	*Isolation method 2 (activated carbon)*: water: 5.07% methanol: 4.67% isopropanol: 5.28%	[41]

Table 2. (Continued)

Sample	Extracted compounds	Conditions tested	Results	Reference
		2) Each extract obtained was adjusted to pH 3.0 with phosphoric acid, after, 40 g/L of activated carbon was added, and magnetically stirred at 60°C for 30 minutes. Then, it was vacuum filtered through celite. CGA were desorbed using ethanol 96%, then dried with anhydrous sodium sulphate, and in a rotary evaporator at 60°C and 120 rpm.		
Green arabica coffee	Caffeine (1,3,7-trimethylxanthine. 4)	Extraction: 20 g of each sample with 90 mL of distilled water were left to reflux for 30 minutes.	Isolation of high purity caffeine (4) from five varieties of green Arabica coffee. The caffeine content in the samples was 1.100 - 1.322 g caffeine (4)/100 g plant product.	[42]
Green coffee beans (*C. arabica* L. cv. Medellin) at different roasting degrees	Total phenolic compounds (TPC) and total flavonoids compounds (TFC)	*Roasting process:* 240°C time of roasting degrees: light: 9 min medium: 11 min dark: 13 min very dark: 16 min	*(TPC)* unroasted: 36.93 mg GAE^1/g light-roasted: 55.00 mg GAE^1/g medium-roasted: 41.49 mg GAE^1/g dark-roasted: 35.95 mg GAE^1/g very dark-roasted: 26.80 mg GAE^1/g	[40]

Sample	Extracted compounds	Conditions tested	Results	Reference
		Extraction conditions: 50 g of each sample were dissolved in 1 L of distilled water and stirred for 5 min at 80°C	*(TFC)* unroasted: 20.41 mg CE^2/g light-roasted: 24.79 mg CE^2/g medium-roasted: 15.40 mg CE^2/g dark-roasted: 9.30 mg CE^2/g very dark-roasted: 4.88 mg CE^2/g	
Mexican green coffee beans (*Coffea arabica*)	Chlorogenic acids	*Sample*: lyophilized green coffee powder *Extraction methods*: 100 g of sample with 500 mL of solvent were magnetically stirred at different conditions in each method, after they were filtered through celite. Water: stirred for 30 min at 80°C. Aqueous methanol (70%): stirred for 24 h at room temperature. Methanol was evaporated in a rotary evaporator. Aqueous isopropanol 60%: stirred for 48 h at room temperature, in the dark. Isopropanol was evaporated in a rotary evaporator. *Isolation methods*: 1) Each extract was added with ammonium sulphate until reach a concentration of 20 g/L. After this, 300 mL phosphoric acid (4%) was added. The aqueous phase was extracted 4 times with 300 mL of ethyl acetate. This phase was dried with 10 g of anhydrous sodium sulphate, filtered and evaporated in a rotary evaporator at 40°C and 120 rpm. 2)	Extraction and isolation yields of chlorogenic acids *Isolation method 1*: water: 5.62% methanol: 5.87% isopropanol: 5.87% *Isolation method 2 (activated carbon)*: water: 5.07% methanol: 4.67% isopropanol: 5.28%	[41]

Table 2. (Continued)

Sample	Extracted compounds	Conditions tested	Results	Reference
		Each extract obtained was adjusted to pH 3.0 with phosphoric acid, after, 40 g/L of activated carbon was added, and magnetically stirred at 60°C for 30 minutes. Then, it was vacuum filtered through celite. CGA were desorbed using ethanol 96%, then dried with anhydrous sodium sulphate, and in a rotary evaporator at 60°C and 120 ppm.		
Green arabica coffee	Caffeine (1,3,7-trimethylxanthine. 4)	Extraction: 20 g of each sample with 90 mL of distilled water were left to reflux for 30 minutes. After the solution was mixed with 12.5 mL of $(CH_3COO)_2$Pb and heated for 5 minutes, it was extracted with 40 mL of chloroform. This phase was washed with KOH and distilled water, then evaporated. The residue was considered the caffeine in the sample.	Isolation of high purity caffeine (4) from five varieties of green Arabica coffee. The caffeine content in the samples was 1.100 - 1.322 g caffeine (4)/100 g plant product.	[42]
Green arabica coffee	Caffeine (1,3,7-trimethylxanthine. 4)	Extraction: 20 g of each sample with 90 mL of distilled water were left to reflux for 30 minutes. After the solution was mixed with 12.5 mL of $(CH_3COO)_2$Pb and heated for 5 minutes, it was extracted with 40 mL of chloroform. This phase was washed with KOH and distilled water, then evaporated. The residue was considered the caffeine in the sample.	Isolation of high purity caffeine (4) from five varieties of green Arabica coffee. The caffeine content in the samples was 1.100 - 1.322 g caffeine (4)/100 g plant product.	[42]

[1]GAE: gallic acid equivalent; [2]CE: catechin equivalent.

Tsukui, Júnior [32] compared Soxhlet extraction with microwave-assisted extraction (MAE) to extract coffee oil fraction of green Arabica coffee beans from Brazil. The oil and diterpene extraction yield was higher using the Soxhlet extraction than the microwave-assisted extraction (MAE). However, using the MAE presented some advantages such as reducing time extraction (10 min *versus* 4 h using Soxhlet extraction) even if the yield was lower; due to this results authors calculated the space–time yield of diterpene content, and observed 6-fold content compared with Soxhlet method.

Phenolic compounds and flavonoids are phytochemical present in the hydrophilic fraction, thus, the content of these compounds during roasting process and coffee beverage preparation are affected by the use of high temperatures. To know more about these changes Cho, Park [40] examined aqueous extractions of coffee beans at different roasting degrees, replicating the coffee brewer preparation. Through this analysis, it was found that coffee beans slightly roasted preserved major phenolic and flavonoid content as well as their antioxidant activity compared to the unroasted coffee bean and the more roasted coffee bean.

Suárez-Quiroz, Alonso Campos [41] extracted and isolated CGAs (3) from Mexican coffee using three different solvents (methanol 70%, isopropanol 60% and water) and two different isolation processes (activated carbon and ethyl acetate). The quantification of the compounds extracted for all different methods did not show significant differences and were in a range of 5.67–5.87% (dry basis). Although the content was similar, the aqueous extraction and the isolation using activated carbon were found to be the best suitable method, because it is cheaper and environmentally friendly, avoiding the use of organic solvents.

7. IDENTIFICATION OF THE MAIN PHYTOCHEMICALS IN COFFEE BEANS

The coffee extracts aqueous fractions include phenolic compounds such as chlorogenic acids (CGAs), phenolic acids, flavonoids and other compounds. CGAs are a complex group that can be purified to obtained a simple compound, Suárez-Quiroz, Alonso Campos [41] made the purification of 5-O-caffeoylquinic acid (neochlorogenic acid. 17) with a silica gel 60 column (25 cm long, 1.6 cm diameter) using as elution solvent a mixture of toluene/ethyl acetate (90:10, v/v), the fractions were dried in a rotary evaporator at 60°C and 120 rpm. After the separation of this compound, or other phenolic compounds, it is important to identify them, the most common method is high performance liquid chromatography (HPLC). Also, 5-O-caffeoylquinic acid (neochlorogenic acid. 17) were identified by high performance liquid chromatography (HPLC) at 276 nm using as mobile phase methanol 70% at 40°C [41]. A reverse-phased C_{18} column (ODS) can be used for the identification of the phenolic compounds, comparing coffee extracts and the retention times at a wavelength of 320 nm, with standards such as CGAs and other phenolic acids. The mobile phase used includes acetic acid/water and acetic acid/acetonitrile/ water [40], other mobile phases can be formic acid and methanol, which separated CGAs at 30°C at 325 nm, and they can be identified through the injection of standards [42].

Caffeine (1,3,7-trimethylxanthine. 4) was extracted and quantitatively identified through the red coloration of the chemical reactions with $AgNO_3$ and $K_2Cr_2O_7$. The identification was developed using thin layer chromatography (TLC) on silica gel sheets, which are revealed by exposing them to iodine vapor, and viewing the brown spots with a source of ultraviolet (UV) light (265 nm). The solvents mixture used was butanol saturated with distilled water and formic acid. The caffeine (4) standard retention factor (Rf) was calculated and then compared to the Rf obtained in the coffee beans extracts[42]. To obtain the content of caffeine (4) and trigonelline (5) in coffee extracts, Dong et al. (2017) used HPLC method in a reversed-phase C_{18} column. The mobile phase included methanol and glacial acetic acid aqueous solution (0.1%). The column worked at 25° C, and the detection was set at 275 nm and 268 nm for caffeine (4) and trigonelline (5), respectively [18].

After the oil extraction from coffee beans, it is necessary to make the transesterification as the first step to identify diterpenes, this procedure includes a saponification of the sample with KOH 2.5 M in methanol at 70°C for 60 min. After this, the residue obtain is cooled and evaporated, and then extracted with methyl *tert*-butyl ether (MTBE) and water, the organic phase is collected after the extract centrifugation, and the aqueous phase is wash with organic solvent to obtain an organic phase, which is washed with citric acid solution until reach an acidic pH. The organic solvent is then evaporated, and the residue is dissolved in methanol [39]. Cafestol (CAF. 1) and kahweol (KAH. 2) can be identified and quantified by HPLC analysis using reversed-phase C_{18} column and standards as reference for the retention time. Araújo and Sandi [36] and Tsukui, Júnior [32] used a mixture of methanol/water as a mobile phase and a wavelength of 220 nm. Belandria, Aparecida de Oliveira [39] used different wavelengths for the identification of cafestol (CAF. 1) (230 nm) and kahweol (KAH. 2) (290 nm). The retention time was 28 min for cafestol (CAF. 1) and 28 min for kahweol (KAH. 2).

8. HEALTH BENEFITS OF COFFEE'S BIOACTIVE COMPOUNDS

A cup of home-made coffee in Canada contains an average of 30 to 175 mg of bioactive compounds (caffeine, CGAs and diterpenes), whereas a cup of United States coffee contains 85 mg for ground roasted coffee, 60 mg for instant coffee and 3 mg for decaffeinated coffee, and the quantity of caffeine in one espresso may reach 200–300 mg [6]. The content of bioactive compounds in green coffee beans oscillated between 1.5-2.5% in dry weight and during the roasting process the content is affected due to its thermic stability [2, 8, 43]. This means that there are not two cups of coffee with the same chemical composition, even when the coffee comes from the same outlet [8, 44]. A daily intake of three to four 8-oz cups of brewed coffee (a total of 400 mg/day of caffeine (1,3,7-trimethylxanthine. 4) or a moderate amount such as 110-345 mg/day of caffeine (4)) for most adults, appears to be associated with a neutral to potentially beneficial effect [2, 44].

The beneficial properties of coffee seem to be dependent on the phenolic acids dose; the most important property of phenolic acids is the protection against the oxidative damage and stress damage such as atherothrombosis and atherosclerotic lesion, improving the development through endothelial protection attenuating oxidative stress, improving the nitric oxide bioavailability, and decreasing the endothelial leukocyte adhesion molecule-1 (E-selectin. ELAM-1), intercellular adhesion molecule-1 (ICAM-1), and vascular cell adhesion molecule-1 (VCAM-1) expression, among others [2, 6, 12, 25, 46].

The chlorogenic acids (CGAs) increase enzymatic activities of glutation *S*-transferase (GST) and suppress the chemical-induced neoplastic transformation of cancer cells [45]. Some studies shown that the consumption of 3-4 cups of decaffeinated coffee has high content of CGAs, reducing in a 30% the risk of type 2 diabetes and diminishing blood glucose levels by the inhibition of glucose-6-phosphatase (G6Pase) activity, affecting hepatic gluconeogenesis with the subsequent delaying on glucose absorption; for all of this chlorogenic acids (CGAs) are an important glucose metabolism regulator, improving insulin functions such as the therapeutic action of an antidiabetic drug metformin (24) (Figure 9), known as the beginner of insulin sensitizer [25]. It has been noticed that short-term coffee consumption might block the effects of adenosine A1 receptor, thus, decreasing glucose uptake in skeletal muscles, which may result in insulin resistance and impaired glucose tolerance. However, long-term and habitual coffee consumption results in normal glucose tolerance improving the insulin sensitivity.

Furthermore, 3-caffeoylquinic acid. (3) and 5-*O*-caffeoylquinic acid (neochlorogenic acid. 5-CQA. 17) are considered a weight-loss supplement due to they have shown an inverse relationship on gaining weight [25]. The possible mechanism proposed for this is the inhibition of glucose absorption and hepatic glucose-6-phosphatase activity, that may alter the secretion of gut peptides, such as glucagon-like peptide-1 (GLP-1) and glucose-dependent insulinotropic polypeptide (GIP), which may play an important role in the short-term regulation of energy intake, regulation of satiety and increase of body fat catabolism [47, 48].

Currently, the studies about brain health have shown that coffee polyphenols, mainly 3-caffeoylquinic acid (3) and 5-*O*-caffeoylquinic acid (neochlorogenic acid. 5-CQA. 17) combined with caffeine, protected against the cognitive deterioration of Alzheimer's disease (AD) in the central nervous system (CNS) [6]. However, epidemiological studies in humans are sparse and controversial, based in a case–control design. In these studies the progression of dementia was reduced in mild cognitive impairment subjects who had a higher concentration of caffeine in blood [48].

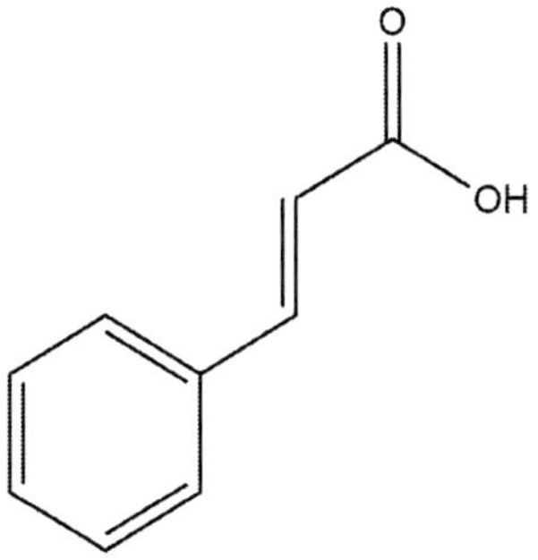

hydroxycinnamic acid (21)

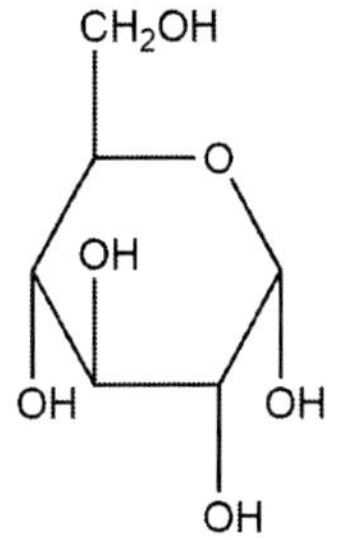

glucose (22)

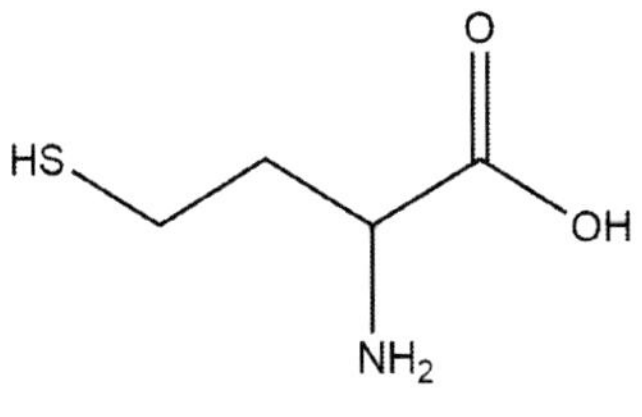

homocysteine (23)

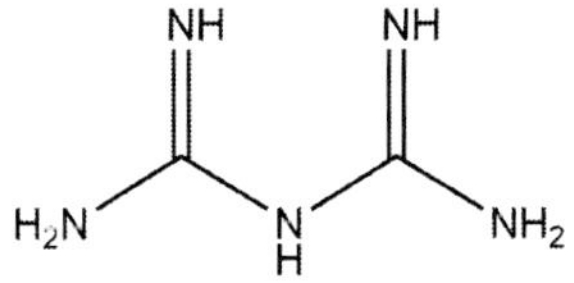

metformin (24)

Figure 9. Structures of some compounds responsible for health benefits hydroxycinnamic acid (21), glucose (22), and homocysteine (23), and metformin (24).

The habitual caffeinated-coffee consumption may protect of the effects of neurodegenerative diseases such as Alzheimer's disease (AD) and

Parkinson's disease (PD). Currently the meta-analysis data found a 33% reduction in the risk to develop Parkinson's disease (PD); although the protection mechanism is still unknown, it has been proposed a link between coffee and glutamate-receptor gene *GRIN2A*, that may modulate the risk of developing Parkinson's disease (PD) in heavy coffee drinkers, using genome wide-based technology [49]. On the other hand, caffeinated coffee has also been associated with the reduction of markers for hepatic cell damage, reduction of chronic liver disease risk, cirrhosis risk reduction, and also reduction of the levels of gamma-glutamyl transferase (GGT), alanine aminotransferase (ALT), and aspartate aminotransferase (AST), which are a liver cell injury and inflammation biomarkers [45].

8.1. Health Bioactive Compounds Found in Coffee

As mention before, coffee consumption has several health benefits, which are attributed to the synergism among bioactive compounds in the brewed coffee [6, 12, 45]. After coffee consumption, each family of bioactive compounds has different absorption mechanisms.

8.1.1. Chlorogenic Acids (CGAs)

Some studies reported that *Coffea Robusta* green beans have higher chlorogenic acids (CGAs) content compared to *Coffea arabica* green beans. A serving of 200 mL of coffee brew provides between 30 and 350 mg of CGAs depending on coffee variety, roasting degree, and brewing methods. After coffee consumption CGAs are hydrolized into caffeic acid (8) and (-)-quinic acid (11), mainly by esterase enzymes and gut microbiota in the gastrointestinal tract. After absorption they are metabolized to glucuronide and sulphate metabolites as the circulating forms in human plasma [6, 25].

The metabolism of CGAs is still unclear, although studies in humans have confirmed that it mainly occurs at two stations (small intestine and colon). The first step is carried out by the active esterase enzymes, which generate the original phenolic acids in both, the small and large bowel.

Absorption in the colon is probably the most quantitatively relevant, representing around two-thirds of the ingested CGAs. The metabolism is carried out by the microbiota, which cleaves the ester bond and provides esterases for further metabolism [6].

8.1.2. Alkaloids

After coffee beverage is prepared, it is known that caffeine (4) has a half-life of approximately 4–6 h. Once coffee beverage is consumed, caffeine (4), absorption occurs primarily in liver; where the cytochrome P450 isoform CYP1A2 accounts for almost 95% of the primary metabolism. The main action of caffeine (4) is to behave as adenosine receptor antagonist (A1, A2A, A2B, and A3) [2, 6], adenosines are presented in most of the tissues, such as central nervous system (CNS), vascular endothelium, heart tissue, liver tissue, adipose tissue, and muscle-conferring cardio-protection, and it attenuates the negative effects of antagonism of the A_1 and A_{2A} adenosine receptors, it is also associated with depression risk reduction. Studies reported that caffeine additional benefit effect is weight control by inducing thermogenesis, increasing the basal metabolic rate (BMR) and fat oxidation in normal-weight subjects [47]. Recent studies also suggest that caffeine may inhibit liver cancer cell proliferation, and reduces the incidence and multiplicity of carcinogen-induced liver cancer in rats [45]. Additionally, coffee may improve asthma symptoms since caffeine (4) is a bronchodilator.

In addition, trigonelline (5) has shown potential therapeutic properties such as hypoglycemic, neuroprotective and anticarcinogenic. Ludwig, Clifford [50] observed that trigonelline (5) is an effective inhibitor of nuclear respiratory factor 2 (Nrf2). Trigonelline (5) is also associate at glucose (22) and lipid metabolism enzyme regulator in diabetic rats (e.g., glucokinase (GK), glucose-6-phosphatase (G6Pase), fatty acid synthase (FAS), and carnitine palmitoyltransferase I (CPT1)); moreover, it has shown properties as a potential antimicrobial agent against *Salmonella enterica* and diminishing the adhesive properties of *Streptococcus mutans* by reducing the bacteria adsorption ability onto saliva-coated hydroxy-apatite beads [50].

8.1.3. Diterpenes

Unfiltered coffee with high quantities of diterpenes have shown a reducing effect of genotoxicity of aflatoxin B1 on cell lines. Kahweol (KAH. 2) and cafestol (CAF. 1) might enhance the detoxification of carcinogens such as *N*-nitrosodimethylamine (NDMA) and 2-amino-1-methyl-6-phenylimidazopyridine (PhIP) by inducing liver enzymes [24].

Studies in animals and cell line models have suggested that cafestol (CAF. 1) and kahweol (KAH. 2) may exert anti-hepatocarcinogenesis effects and may reduce mRNA and protein levels of cytochrome P450 (CYP) isoenzymes 1A1, 1A2, 3A2, 2B1, 2B2, and 2C11, along with the reduced formation of aflatoxin B1 (AFB1)-DNA adducts [45].

8.1.4. Other Compounds

During coffee brew preparation, some polysaccharides such as arabinogalactans, galactomannans, mannans, galactose, arabinose and some of their derivatives are solubilized. At the high temperatures used to prepare coffee beverage, copolymerization reactions occurred; leading to Maillard reaction, increasing melanoidin content. Melanoidins might inhibit matrix metalloproteinases (MMPs; endo-peptidases), which play a central role in tumor growing and metastasis, as well protect against colon cancer, since matrix metalloproteinase 1 (MMP-1), matrix metalloproteinase 2 (MMP-2), and matrix metalloproteinase 9 (MMP-9) are thought to be involved in the pathogenesis of colon cancer, and modulate the bacterial colon population (modulate colonic microflora) [50, 51 52].

Melanoidin health implications are derived from their antioxidant activity that has been associated with protective effects on cultured hepatoma HepG2 human cells, decreasing glutathione peroxidase (GPx) and glutathione reductase (GR) activity against oxidative stress, and inhibition of the lipoxidation involved in the development of atherosclerosis and other diseases [50, 51]. In addition, melanoidins promote dental caries protection, related to its ability to adhere to the tooth surface and form a biofilm or by affecting *Streptococcus mutans* and the inhibition of biofilm formation on microtiter plates [50, 51]. Moreover, melanoidin antimicrobial activity is attributed to their metal-chelating properties, and three different action mechanisms have been proposed: (i) bacteriostatic activity mediated by iron chelation from the culture medium; (ii) chelating the siderophore-Fe^{3+} complex in bacteria able to produce siderophores for iron acquisition, decreasing the virulence of the bacteria; and (iii) bactericidal activity at high concentrations by removing Mg^{2+} ions from the outer membrane, thus, promoting the disruption of the cell membrane [50].

CONCLUSION

Coffee is an important beverage, mainly obtained from two varieties of coffee beans (robusta and arabica), and it is consumed worldwide due to its sensorial properties such as great taste and odor. Besides its organoleptic features, coffee contains many proven bioactive molecules with synergic properties such as caffeine (1,3,7-trimethylxanthine. 4), chlorogenic acid (CGA. 3-caffeoylquinic acid. 3), diterpenes and alkaloids. These coffee phytochemicals exert important biological activities such as anticancer, antimicrobial, antioxidant, and others. Due to the importance of this phytochemicals, different analytical methods have been proposed to add quality assurance and safe guarding consumer satisfaction and safety.

REFERENCES

[1] Bae JH, Park JH, Im SS, Song DK. Coffee and health. *Integrative Medicine Research* 3(4), 189-191, 2014.

[2] O'Keefe JH, Bhatti SK, Patil HR, DiNicolantonio JJ, Lucan SC, Lavie CJ. Effects of habitual coffee consumption on cardiometabolic disease, cardiovascular health, and all-cause mortality. *J Am Coll Cardiol* 62(12), 1043-1051, 2013.

[3] Organization IC. *Trade Statistics Tables 2017* [Available from: http://www.ico.org/trade_statistics.asp.].

[4] Higdon JV, Frei B. Coffee and health: a review of recent human research. *Crit Rev Food Sci Nutr* 46(2), 101-123, 2006.
[5] Borrell B. Plant biotechnology: make it a decaf. *Nature* 483(7389), 264-266, 2012.
[6] Cano-Marquina A, Tarín JJ, Cano A. The impact of coffee on health. *Maturitas* 75(1), 7-21, 2013.
[7] Akash MS, Rehman K, Chen S. Effects of coffee on type 2 diabetes mellitus. *Nutrition* 30(7-8), 755-763, 2014.
[8] Gonzalez de Mejia E, Ramirez-Mares MV. Impact of caffeine and coffee on our health. *Trends Endocrinol Metab* 25(10), 489-492, 2014.
[9] Lopez-Garcia E, Guallar-Castillon P, Leon-Muñoz L, Graciani A, Rodriguez-Artalejo F. Coffee consumption and health-related quality of life. *Clin Nutr* 33(1), 143-149, 2014.
[10] Pourshahidi LK, Navarini L, Petracco M, Strain JJ. A comprehensive overview of the risks and benefits of coffee consumption. *Comprehensive Reviews in Food Science and Food Safety* 15(4), 671-684, 2016.
[11] Gupta RC. *Nutraceuticals: efficacy, safety and toxicity*. Academic Press, 2016.
[12] Miranda AM, Steluti J, Fisberg RM, Marchioni DM. Association between coffee consumption and its polyphenols with cardiovascular risk factors: A population-based study. *Nutrients* 2017 Mar 14, 9(3). pii: E276. doi: 10.3390/nu9030276.
[13] Crozier A, Ashihara H. *Teas, cocoa and coffee: plant secondary metabolites and health*. John Wiley & Sons, 2011.
[14] Homan DJ, Mobarhan S. Coffee: good, bad, or just fun? A critical review of coffee's effects on liver enzymes. *Nutr Rev* 64(1), 43-46, 2006.
[15] Gómez-Ruiz JA, Leake DS, Ames JM. In vitro antioxidant activity of coffee compounds and their metabolites. *J Agric Food Chem* 55(17), 6962-6969, 2007.
[16] Williamson G. *Coffee and health. Teas, Cocoa and Coffee*. pp169-192. Wiley-Blackwell, 2011.

[17] Esquivel P, Jiménez VM. Functional properties of coffee and coffee by-products. *Food Research International* 46(2), 488-495, 2012.
[18] Dong W, Hu R, Chu Z, Zhao J, Tan L. Effect of different drying techniques on bioactive components, fatty acid composition, and volatile profile of robusta coffee beans. *Food Chemistry* 234, 121-130, 2017.
[19] Aguiar J, Estevinho BN, Santos L. Microencapsulation of natural antioxidants for food application–The specific case of coffee antioxidants–A review. *Trends in Food Science & Technology* 58, 21-39, 2016.
[20] Bicho NC, Lidon FC, Ramalho JC, Leitao AE. Quality assessment of arabica and robusta green and roasted coffees-A review. *Emirates Journal of Food and Agriculture* 25(12), 945-950, 2013.
[21] Affonso RC, Voytena AP, Fanan S, Pitz H, Coelho DS, Horstmann AL, Pereira A, Uarrota VG, Hillmann MC, Varela LA, Ribeiro-do-Valle RM, Maraschin M. Phytochemical composition, antioxidant activity, and the effect of the aqueous extract of coffee (*Coffea arabica* L.) bean residual press cake on the skin wound healing. *Oxid Med Cell Longev* 2016;2016:1923754. Epub 2016 Nov 14.
[22] de Roos B, van der Weg G, Urgert R, van de Bovenkamp P, Charrier A, Katan MB. Levels of cafestol, kahweol, and related diterpenoids in wild species of the coffee plant coffea. *J Agric Food Chem* 45(8), 3065-3069, 1997.
[23] Moreno J, Peinado R. *Enological chemistry*. Academic Press, 2012.
[24] Tuan PA, Kwon DY, Lee S, Arasu MV, Al-Dhabi NA, Park NII, Park SN. Enhancement of chlorogenic acid production in hairy roots of *Platycodon grandiflorum* by over-expression of an *Arabidopsis thaliana* transcription factor AtPAP1. *Int J Mol Sci* 15(8), 14743–14752, 2014.
[25] Tajik N, Tajik M, Mack I, Enck P. The potential effects of chlorogenic acid, the main phenolic components in coffee, on health: a comprehensive review of the literature. *Eur J Nutr* 2017 Apr 8. doi: 10.1007/s00394-017-1379-1.

[26] Liang N, Kitts DD. Antioxidant property of coffee components: assessment of methods that define mechanisms of action. *Molecules* 19(11), 19180-19208, 2014.
[27] Monteiro JP, Alves MG, Oliveira PF, Silva BM. Structure-bioactivity relationships of methylxanthines: Trying to make sense of all the promises and the drawbacks. *Molecules* 2016 Jul 27;21(8). pii: E974. doi: 10.3390/molecules21080974.
[28] Lean MEJ, Ashihara H, Clifford MN, Crozier A. *Purine alkaloids: A focus on caffeine and related compounds in beverages. Teas, cocoa and coffee.* pp5-44. Wiley-Blackwell, 2011.
[29] Godos J, Pluchinotta FR, Marventano S, Buscemi S, Li Volti G, Galvano F, Grosso G. Coffee components and cardiovascular risk: beneficial and detrimental effects. *Int J Food Sci Nutr* 65(8), 925-936, 2014.
[30] Ivamoto ST, Sakuray LM, Ferreira LP, Kitzberger CS, Scholz MB, Pot D, Leroy T, Vieira LG, Domingues DS, Pereira LF. Diterpenes biochemical profile and transcriptional analysis of cytochrome P450s genes in leaves, roots, flowers, and during *Coffea arabica* L. fruit development. *Plant Physiol Biochem* 111, 340-347, 2017.
[31] Scharnhop H, Winterhalter P. Isolation of coffee diterpenes by means of high-speed countercurrent chromatography. *Journal of Food Composition and Analysis* 22(3), 233-237, 2009.
[32] Tsukui A, Santos Júnior HM, Oigman SS, de Souza RO, Bizzo HR, Rezende CM. Microwave-assisted extraction of green coffee oil and quantification of diterpenes by HPLC. *Food Chem* 164, 266-271, 2014.
[33] Jeszka-Skowron M, Zgoła-Grześkowiak A, Grześkowiak T. Analytical methods applied for the characterization and the determination of bioactive compounds in coffee. *European Food Research and Technology* 240(1), 19-31, 2015.
[34] Opitz SEW, Goodman BA, Keller M, Smrke S, Wellinger M, Schenker S, Yeretzian C. Coffee brews by coupling on-line ABTS assay to high performance size exclusion. chromatography. *Phytochem Anal* 28(2), 106-114, 2017.

[35] Dulsat-Serra N, Quintanilla-Casas B, Vichi S. Volatile thiols in coffee: A review on their formation, degradation, assessment and influence on coffee sensory quality. *Food Research International* 89(2), 982-988, 2016.
[36] Araújo JMA, Sandi D. Extraction of coffee diterpenes and coffee oil using supercritical carbon dioxide. *Food Chemistry* 101(3), 1087-1094, 2007.
[37] Da Porto C, Porretto E, Decorti D. Comparison of ultrasound-assisted extraction with conventional extraction methods of oil and polyphenols from grape (*Vitis vinifera* L.) seeds. *Ultrasonics Sonochemistry* 20(4), 1076-1080, 2013.
[38] Ghasemi E, Raofie F, Najafi NM. Application of response surface methodology and central composite design for the optimisation of supercritical fluid extraction of essential oils from *Myrtus communis* L. leaves. *Food Chemistry* 126(3), 1449-1453, 2011.
[39] Belandria V, Aparecida de Oliveira PM, Chartier A, Rabi JA, de Oliveira AL, Bostyn S. Pressurized-fluid extraction of cafestol and kahweol diterpenes from green coffee. *Innovative Food Science & Emerging Technologies* 37(Part A), 145-152, 2016.
[40] Cho AR, Park KW, Kim KM, Kim SY, Han J. Influence of roasting conditions on the antioxidant characteristics of Colombian coffee (*Coffea arabica* L.) Beans. *Journal of Food Biochemistry* 38(3), 271-280, 2014.
[41] Suárez-Quiroz ML, Alonso Campos A, Valerio Alfaro G, González-Ríos O, Villeneuve P, Figueroa-Espinoza MC. Isolation of green coffee chlorogenic acids using activated carbon. *Journal of Food Composition and Analysis* 33(1), 55-58, 2014.
[42] Patriche S, Boboc M, Leah V, Dinica R-M. Extraction and evaluation of bioactive compounds with antioxidant potential from green arabica coffee extract. *Annals of the University of Dunarea de Jos of Galati. Fascicle VI Food Technology* 39(2), 88-95, 2015.
[43] Farah A. *Coffee constituents. Coffee.* pp 21-58. Wiley-Blackwell, 2012.

[44] Rodrigues NP, Bragagnolo N. Identification and quantification of bioactive compounds in coffee brews by HPLC–DAD–MS". *Journal of Food Composition and Analysis* 32(2), 105-115, 2013.

[45] Johnson S, Koh WP, Wang R, Govindarajan S, Yu MC, Yuan JM. Coffee consumption and reduced risk of hepatocellular carcinoma: findings from the Singapore Chinese Health Study. *Cancer Causes Control* 22(3), 503-510, 2011.

[46] Rio DD, Rodriguez-Mateos A, Spencer JPE, Tognolini M, Borges G, Crozier A. Dietary (poly) phenolics in human health: structures, bioavailability, and evidence of protective effects against chronic diseases. *Antioxid Redox Signal* 18(14), 1818–1892, 2013.

[47] Bakuradze T, Parra GAM, Riedel A, Somoza V, Lang R, Dieminger N, Hofmann T, Winkler S, Hassmann U, Marko D, Schipp D, Raedle J, Byto G, Lantz I, Stiebitz H, Richling E. Four-week coffee consumption affects energy intake, satiety regulation, body fat, and protects DNA integrity. *Food Research International* 63(Part C):420-427, 2014.

[48] Cao C, Loewenstein DA, Lin X, Zhang C, Wang L, Duara R, Wu Y, Giannini A, Bai G, Cai J, Greig M, Schofield E, Ashok R, Small B, Potter H, Arendash GW. High blood caffeine levels in MCI linked to lack of progression to dementia. *J Alzheimers Dis* 30(3), 559-572, 2012.

[49] Hamza TH, Chen H, Hill-Burns EM, Rhodes SL, Montimurro J, Kay DM, Tenesa A, Kusel VI, Sheehan P, Eaaswarkhanth M, Yearout D, Samii A, Roberts JW, Agarwal P, Bordelon Y, Park Y, Wang L, Gao J, Vance JM, Kendler KS, Bacanu SA, Scott WK, Ritz B, Nutt J, Factor SA, Zabetian CP, Payami H. Genome-wide gene-environment study identifies glutamate receptor gene GRIN2A as a Parkinson's disease modifier gene via interaction with coffee. *PLoS Genet* 2011 Aug;7(8):e1002237. doi: 10.1371/journal.pgen.1002237. Epub 2011 Aug 18.

[50] Ludwig IA, Clifford MN, Lean ME, Ashihara H, Crozier A. Coffee: biochemistry and potential impact on health. *Food Funct* 5(8), 1695-1717, 2014.

[51] Moreira AS, Nunes FM, Domingues MR, Coimbra MA. Coffee melanoidins: structures, mechanisms of formation and potential health impacts. *Food Funct* 3(9), 903-915, 2012.

[52] Nunes FM, Coimbra MA. Chemical characterization of the high molecular weight material extracted with hot water from green and roasted arabica coffee. *J Agric Food Chem* 49(4), 1773-1782, 2001.

In: Occurrences, Structure, Biosynthesis ... ISBN: 978-1-53614-141-2
Editor: Noboru Motohashi © 2018 Nova Science Publishers, Inc.

Chapter 3

NATURE-INSPIRED PHYTOCHEMICALS AND THE PHARMACOLOGICAL ACTIVITIES OF HERBAL PLANTS OF THE ANACARDIACEAE FAMILY AND *SEMECARPUS ANACARDIUM* L.F.

Vustelamuri Padmavathi[1,], Bhattiprolu Kesava Rao[1,†] and Noboru Motohashi[2,+]*

[1]Department of Chemistry, University College of Sciences, Acharya Nagarjuna University, Nagarjunanagar, Andhra Pradesh, India
[2]Meiji Pharmaceutical University, Noshio, Kiyose-shi, Tokyo, Japan

---

[*] Email: Padma1202@gmail.com
[†] Email: krbhattiprolu@gmail.com. Chairman-BOS, Department of Chemistry, University College of Sciences, Acharya Nagarjuna University, Nagarjunanagar- 522 510, Guntur District, Andhra Pradesh, India.
[+] Email: noborumotohashi@jcom.home.ne.jp

ABSTRACT

In the 21St century, herbal drugs are contributing much to human health. In our tradition, especially natural medicine improves the inner immune system of the human body and no adverse effects could be observed. Hence, the herbal drug acts more effectively than the modern medicine. In our survey, we found that, the *Anacardiaceae* family has 83 genera and 860 species existing as trees, shrubs and vines. Out of these, the commonly known cashew family or sumac family has tremendous commercial importance. Due to the presence of several anticancer drugs isolated from *Anacardiaceae* family, we have selected *Semecarpus anacardium* L.f. for its high medicinal value in Ayurveda and Siddha systems and isolated several active constituents like phenolic lipids, flavonoids and biflavonoids. Then, this review tried to describe separately the properties and pharmacological effects of *Anacardiaceae* family and *Semecarpus anacardium* L.f. in three parts of PART 1: Natural Products and Herbal Plants, PART 2: Anacardiaceae family, and PART 3: *Semecarpus anacardium* L.f..

Keywords: anacardiaceae family, *Semecarpus anacardium* L.f., phenolic lipids, flavonoids,biflavonoids and pharmacological activities

CHEMICAL CONSTITUENTS

- artemisinin (**1**)
- taxol (**2**)
- camptothecin (**3**)
- morphine (**4**)
- strychnine (**5**)
- quinine (**6**)
- codeine (**7**)
- digitoxin (**8**)
- atropine (**9**)
- hyoscine (**10**)
- penicillin (**11**)
- vitamin B$_{12}$ (**12**)

- podophyllin (**13**)
- etoposide (**14**)
- caffeine (**15**)
- Atropa belladonna (**16**)
- nicotine (**17**)
- reserpine (**18**)
- yakuchinone A (1-[4'-hydroxy-3'-methoxypheny1]-7-pheny1-3-heptanone (**19**)
- yakuchinone B (1-[4'-hydroxy-3'-methoxypheny1]-7-phenylhept-1-en-3-one (**20**)
- sesquiterpene (**21**)
- lactone parthenolide (**22**)
- parthenolide (**23**)
- thyme oil (**24**)
- thymol (**25**)
- carvacrol (**26**)
- γ-terpinene (**27**)
- myrcene (**28**)
- linalool (**29**)
- *p*-cymene (**30**)
- limonene (**31**)
- 1,8-cineole (**32**)
- α-pinene (**33**)
- rosmarinic acid (**34**)
- 3,4-dihydroxyphenyllacetic acid (**35**)
- amentoflavone (**36**)
- anastatin A (**37**)
- anastatin B (**38**)
- azorellanol (**39**)
- 13-hydroxy-7-oxoazorellane (**40**)
- 7-deacetylazorellanol (**41**)
- plumbagin (5-hydroxy-2-methyl-1,4-naphthoquinone (**42**)

- β-pinene (**43**)
- terpinolene (**44**)
- terpinen-4-ol (**45**)
- sabinene (**46**)
- *cis*-savinene hydrate (**47**)
- cotyledon (**48**)
- pinene (**49**)
- sabinene hydrate (**50**)
- 3-dimethyl-2-oxabicyclo[2,2,2]-octane (**51**)
- 1,3,3-trimethyl-2-oxabicyclo[2,2,2]octane (**52**)
- (*E*,1*R*,9*S*)-4,11,11-trimethyl-8-methylenebicyclo[7,2,0]undec-4-ene (**53**)
- (*Z*)-3-methyl-2-(pent-2-ynyl)-cyclopent-2-enone (**54**)
- caripill (**55**)
- acarbose (**56**)
- epirubicin (**57**)
- cefprozil (**58**)
- octalactin A (**59**)
- spongouridine (**60**)
- spongothymidine (**61**)
- (+)-discodermolide (**62**)
- captopril (**63**)
- epibatidine (**64**)
- indole alkaloids (**65**)
- vincristine (R = Me(methyl)) (**66**)
- vinblastine (R = CHO) (**67**)
- ajmalicine (**68**)
- podophyllotoxin (PPT. **69**)
- teniposide (**70**)
- paclitaxel (**71**)
- poelophyllotoin (**72**)
- podophylloforim (**73**)

- epothilones (**74**)
- dixabepilone (**75**)
- topotecan (**76**)
- irinotecan (**77**)
- (3-((8Z,11E,13E)-pentadeca-8,11,13-trien-1-yl)benzene-1,2-diol (**78**)
- (Z)-3-(pentadec-8-en-1-yl)benzene-1,2-diol (**79**)
- 3-((8Z,11Z,13Z)-pentadeca-8,11,13-trien-1-yl)benzene-1,2-diol (**80**)
- 3-((10Z,13Z)-heptadeca-10,13,16-trien-1-yl)benzene-1,2-diol (**81**)
- (Z)-2-hydroxy-3-(octadec-8-en-1-yl)benzoic acid (**82**)
- 2-heptadecyl-6-hydroxybenzoic acid (**83**)
- 2-hydroxy-3-methyl-6-((8Z,11Z)-pentadeca-8,11-dienyl)benzoic acid (**84**)
- 3-((8Z,11Z,14Z)-hexadeca-8,11,14-trien-1-yl)phenol (**85**)
- 2-pentadecylbenzoic acid (**86**)
- (Z)-5-(hexadec-8-en-1-yl)benzene-1,3-diol (**87**)
- (Z)-2-(heptadec-8-en-1-yl)benzene-1,4-diol (**88**)
- (4S)-5-((Z)-heptadec-8-en-1-yl)-4,5-dihydroxycyclohex-2-enone (**89**)
- (Z)-3-(heptadec-7-en-1-yl)phenol (**90**)
- (S,Z)-3-(16-phenylhexadec-12-en-1-yl)cyclohexa-1,5-diene-1,3-diol (**91**)
- 3-(14-phenyltetradecyl)phenol (**92**)
- 5-((4E,7E)-trideca-4,7-dien-1-yl)benzene-1,3-diol (**93**)
- 5-((4Z,7Z,10Z)-trideca-4,7,10-trien-1-yl)benzene-1,3-diol (**94**)
- 5-((4Z,7Z,10E)-trideca-4,7,10-trien-1-yl)benzene-1,3-diol (**95**)
- (Z)-5-(dodec-3-en-1-yl)benzene-1,3-diol (**96**)
- 3-(pentadeca-7,10-dien-1-l)benzene-1,2-diol (**97**)
- (Z)-3-(pentadec-10-en-1-yl)benzene-1,2-diol (**98**)
- (4S,5R)-5-((E)-heptadec-14-en-1-yl)-5-hydroxy-4-methylcyclohex-2-enone (**99**)

- (4S,5R)-5-hydroxy-4 methyl-5-((E)-nonadec-16-en-1-yl)cyclohex-2-enone (**100**)
- I-4^1,II-3^1,4^1,I-5,II5,I7-hexahydroxy[I3,II8]biflavonone (**101**)
- I-4^1,II-4^1,I-5,II-5,I-7,II-7-hexahydroxy[I-3,II-8]biflavonone(3^1,8-binaringenin) (**102**)
- I-4^1,II-4^1,I-7,II-7-tetrahexahydroxy [I-3,II-8]biflavonone(3^1,8-billiquiritigenin) (**103**)
- (E)-3-(pentadec-8-enyl)benzene-1,2-diol (**104**)
- 3-((7E,10E)-pentaca-7,10-dienyl)benzene-1,2-diol (**105**)
- tetrahydroamentoflavonone (**106**)
- agathisflavone (**107**)
- jeediflavone (**108**)
- semecarpuflavonone (**109**)
- galluflavonone (**110**)
- nallaflavonone (**111**)
- quercetin (**112**)
- miricetina (myricetin. **113**)
- rhuscorlaria (**114**)
- cotinuscoggygria (**115**)
- rhusalatal (**116**)
- rhussuccedanea (**117**)
- rhussalicifolia (**118**)
- kaempferol (**119**)
- rhuspyroides (**120**)
- kaempferide (**121**)
- 8,8-dimethyl-10-(3-methylbut-2-en-1-yl)-2-phenyl-2,3,7,8-tetrahydropyrano[3,2-g]chromen-4(6H)-one (**122**)
- tetrahydrohinokiflavone (**123**)
- 7-O-methylpelargonidin glycoside (**124**)
- 7-O-methylcyanidin glycoside (**125**)
- rhamnetin (**126**)
- gallocatechin (**127**)
- bhilwanol (**128**)

- anacardic acid (**129**)
- 1-pentadeca-7,10-dienyl-1,3-dihydroxybenzene (**130**)
- 1-pentadeca-8-enyl-2,3-dihydroxybenzene (**131**)
- tetrahydrorobustaflavone (**132**)
- semecarpetin (**133**)
- anacardoflavonone (**134**)
- anacardoside (**135**)
- 3-pentadecyl benzene (**136**)
- (Z)-3-(pentadec-8-enyl)benz-1,2-diol (**137**)
- 3-((8Z,11Z)-pentadeca-8,11-dienyl) benzene-1,2-diol (**138**)
- 3,4,2^1,4^1-tetrahydroxychalcone (**139**)
- 7,3^1,4^1-trihydroxyflavone (**140**)

1. NATURAL PRODUCTS AND HERBAL PLANTS

1.1. Introduction

It is well known that, India is extremely rich in medicinal plant diversity, distributed in various geographical and environmental conditions and associated with much tribal and folklore knowledge, the people who are living in the remote villages and forests, totally dependent upon the medicinal plants for the survival of their ethnic communities. Even today, large number of Indian population relies on traditional herbal medicine and has attracted considerable global interest in recent years [1]. For thousands of years, countries around the world have used herbal plants to treat illness and maintained their health. We found that, medicinal plants represent precious resources from which bioactive compounds can be isolated and developed into valuable therapeutic agents, with the advent of modern drug discovery technologies such as medicinal chemistry, forensic chemistry, combinatorial chemistry and other drug screening platforms. It has been observed that a wide spectrum of bioassay can be employed for the detection of bioactivity in extracts, fractions, as well as purified compounds of herbal origin.

The medicinal chemistry mainly depends upon the history and medicinal usage information received from Ayurveda medicine, Siddha medicine, and tribal folklore medicine. Basing on this, the structural determination and screening for biological activity of natural products were developed. By different biogenetic pathways, plants generally synthesize novel and surprising new structures. The biological activities of these natural products vary in very diverse. The development and discovery is a continuous process of inventing new therapeutic agents. There is a tremendous need to discover and invent new agents is genuine and necessary for all kinds of living beings in order to survive from the natural environmental and in borne diseases.

According to the World Health Organization (WHO), most of these drugs were developed because of their use in traditional medicine. Recent WHO studies indicates that over 30 percent of the world plant species have at one time or another been used for medicinal-purposes. Of the 2,50,000 higher plant species on Earth, more than 80,000 species have medicinal importance. Although traditional medicine is widespread throughout the world; it is an integral part of each individual culture. For hundreds or even thousands, its practice is used mainly on traditional belief handed down from generation to generation. Oldest of such literature is the Ayurvedic text *Charaka-samhita*. India's use of plants for health care also dates back close to 5000 years. About 8000 herbal remedies have been confirmed in Ayurveda, which is still in use in many dispensaries today.

Now it is possible to rapidly build up extensive libraries of certain classes of organic compounds by the method of combinatorial chemistry.

In the early 19th century, the structure determination and synthesis of natural products have gained natural attention, the screening of medicinal or poisonous plants of tropical forests, Marine flora and fauna, soil samples, fungi and microbes produce new drugs or lead structures and these give more potent synthetic and semi-synthetic products [2]. In India, about 20% of the drugs available are from plant-derived and are particularly useful to treat chronic diseases of liver, kidney and skin.

People prefer Ayurvedic or herbal medicines, as they have practically no adverse effects.

Now, there is a great demand and stress from the people and various governments including pharmaceutical industries for the isolation of active chemical ingredients from the natural herbal plant species and also for the development of novel chemical synthetic strategies for their availability in large quantities to satisfy the immediate needs of the suffering people. The ecological awareness and an increased demand for non-classical therapies may be invoked as the main reasons for this renewal.

Artemisinin (**1**), taxol (**2**) and camptothecin (**3**) (Figure 1) are examples of natural products that are used clinically. Several natural products isolated from plants used in traditional medicine have potent anti-plasmodial action *in vitro* and represent potential sources of new anti-malarial drugs and these were studied by Phillipson and co-workers [3]. There are about 121 drugs of known structures-extracted from higher plants that are still used globally in allopathic medicine.

In addition to above, a good number of novel plant-derivative substances have entered into Western drug markets and clinical plant based researches as made a rewarding progress in all important fields, particularly in anti-cancer and anti-malarial therapies as reported by De Smet PA [4].

Naturally occurring medicinal sources, based on their source, are divided into 4 types as follows:

I. Microbial world: *e.g.*, cephalosporins
II. Plant sources: *e.g.*, paclitaxel
III. Marine sources: *e.g.*, discodermalide
IV. Animal source: *e.g.*, epibatidine

These 121 plant-derived drugs are produced commercially from less than 90 species of higher plants. It is estimated that 80% of anti-tumor and anti-infectious drugs already in the market or under clinical trial are of natural origin obtained directly or indirectly [5, 6].

The first isolated compound is an alkaloid which includes morphine (**4**), strychnine (**5**), quinine (**6**) (Figure 1) *etc*. The early 19th century marked a new era in the use of medicinal plants and the beginning of modern medicinal plant research. Equally, the efficacy of a number of phytopharmaceutical preparations, such as aloe, ginko, garlic or valerian, has been demonstrated by studies that applied the same scientific standards as for synthetic drugs. As a result of all, there is an enormous busy market for crude herbal medicines in addition to purified plant-derived drugs [7].

Prior to World War II, a series of natural products isolated from higher plants became clinical agents and a number are still in use today. Quinine (**6**) from cinchona bark, morphine (**4**) (*Papaver somniferum*) and codeine (**7**) (Figure 1) from the latex of the opium poppy, digoxin (**8**) (Figure 1) from digitalis leaves; atropine (**9**) (Figure 1) (*Atropa belladonna)* and hyoscine (**10**) (Figure 1) from species of the Solanaceae continue to be in clinical use. Recently, natural products are attracting the many of pharmaceutical industry. Large pharmaceutical companies such as Merck, CIBA, Glaxo, Bochringer and Syntex, now have specific departments dedicated to the study of new drugs from natural sources. Glaxocompany is well known for the production of drugs in bulk which are identified as drugs from natural products, *e.g.*: penicillin (**11**), vitamin B$_{12}$ (**12**) (Figure 1) *etc*. and it is giving much importance to further natural products research.

Taxol (**2**) is obtained from the bark of the Western Pacific yew, *Taxus brevifolia*. The isolation and structure determination of taxol (**2**) followed on from experiments that showed that a crude extract was active against cancer cells in laboratory tests. The resin podophyllin (**13**) (Figure 1) obtained from the root of the mayapple, *Podophyllum peltatum*, is toxic and is used clinically to remove warts. The major constituent of the resin is the lignin podophyllotoxin (PPT) which inhibits cell division.

The clinical applications of taxol (**2**), etoposide (**14**) (Figure 1) and artemisinin (**1**) have helped to revive an interest in higher plants as sources of new drugs. Though there is a considerable development in medicinal field, there still remains an urgent need to develop new clinical drugs for numerous diseases which result from the malfunction of the central nervous systems (CNS), *e.g.*, Alzheimer's disease (AD) and Parkinson's disease (PD), epilepsy, migraine, pain, schizophrenia, sleeping disorders, *etc*. Natural products already have a proven track record for CNS activities, *e.g.*, caffeine (**15**), Atropa belladonna (**16**), morphine (**4**), nicotine (**17**), reserpine (**18**) (Figure 1) and it is possible to further such drugs still to be found from nature [8].

According to the World Health Organization (WHO), herbal medicines serve the health needs of about 80% of the world's population [9] especially formillions of people in the vast rural areas of developing countries. Meanwhile, consumers in developed countries are becoming disillusioned with modern health care and are seeking alternatives. The recent resurgence of plant remedies results from several factors as follows:

1) The effectiveness of plant medicines
2) The side effect of most modern drugs
3) The development of science and technology

The latest advances in biotechnology has contributed greatly in the production of commercially important compounds, in plants or plant cell culture, or even to produce completely new compounds. Culturing of plant cells can produce many valuable metabolites including novel medicinal agents and recombinant products. In combination with synthetic chemistry, methodology also affords an attractive route to the synthesis of complex natural products and related compounds of industrial importance as reported by Kumar and co-workers [10].

Many scientific methods of analysis have been developed for the investigation of the constituents and biological activities of medicinal plants. The chromatographic (*e.g.*, TLC, GLC, HPLC), spectroscopic (*e.g.*, UV, IR, ^{1}H and ^{13}C-NMR, MS) and biological (*e.g.*, anticancer, anti-inflammatory, immune stimulant, anti-protozoal) techniques utilized for medicinal plant research are reviewed from time to time. The contribution that advances in scientific methodology have made to our understanding of the actions of some herbal medicines (*e.g.*, Echinacea, ginkgo, St. John's wort, cannabis), as well as to ethnopharmacology and biotechnology are revived in scientific journals. Plants have provided many medicinal drugs in the past and remains as a potential source of novel therapeutic agents. Despite all of the powerful analytical techniques available, the majorities of plant species has not been investigated chemically or biologically in any great details and even well-known medicinal plants require medicinal further clinical study as opined by Phillipson and co-workers [11].

According to Kang and co-workers [12] certain medicinal plants contain anti-inflammatory and anti-oxidative substances that can exert chemo-preventive effects. The methanol extract of *Alpinia oxyphylla* Miguel (Zingiberaceae) inhibits tumor promotion in mouse skin. Two major di-aryl-heptanoids named yakuchinone A (1–[4'-hydroxy-3'-methoxyphenyl]-7-phenyl-3-heptanone) **(19)** and yakuchinone B (1-[4'-hydroxy-3'-methoxyphenyl]-7-phenylhept-1-en-3-one) **(20)** (Figure 1) have been isolated from this medicinal plant. Both compounds have strong inhibitory effects on the synthesis of prostaglandins and leukotrienes *in vitro*. The sesquiterpene **(21)** and lactone parthenolide **(22)** (Figure 1), the principal active component in *Parthenium histerephorus*, have been used conventionally to treat migraines, inflammation, and tumors.

Finally, parthenolide **(23)** (Figure 1) and its derivatives may be useful chemotherapeutic agents to treat these invasive cancers. This was reported by Wen and co-workers [13].

Youdim and co-workers [14] reported that antioxidants minimize the oxidation of lipid components in cell membranes by scavenging free radicals. However, imbalance- between free radical production and removal tends to increase with age causing progressive damage. For the food industry, it is of considerable interest to delay the auto-oxidation of food lipids, which is the cause of reduction in food quality, affecting color, taste, nutritive value, and functionality.

A general orientation towards the use of natural compounds has stimulated research into the potential use of aromatic and medicinal plants as possible antioxidant replacements. This study characterized the antioxidant and pro-oxidant properties of thyme oil and a number of its components. The major components identified in thyme oil were found to inhibit ferric-ion-stimulated lipid peroxidation of rat brain homogenates, although none was as effective as the whole oil. The order of antioxidant activity was; thyme oil (**24**)> thymol (**25**) > carvacrol (**26**)> γ-terpinene (**27**) > myrcene (**28**) > linalool (**29**) > *p*-cymene (**30**) > limonene (**31**) > 1,8-cineole (**32**) > α-pinene (**33**) (Figure 1).

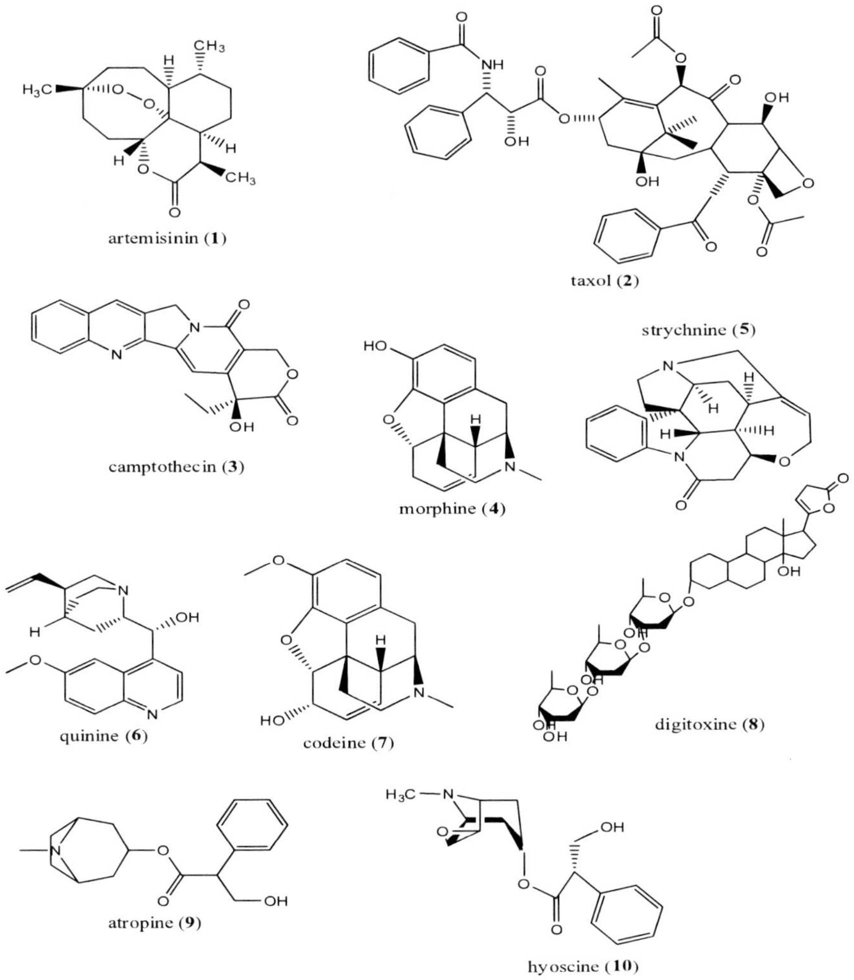

Figure 1. (Continued).

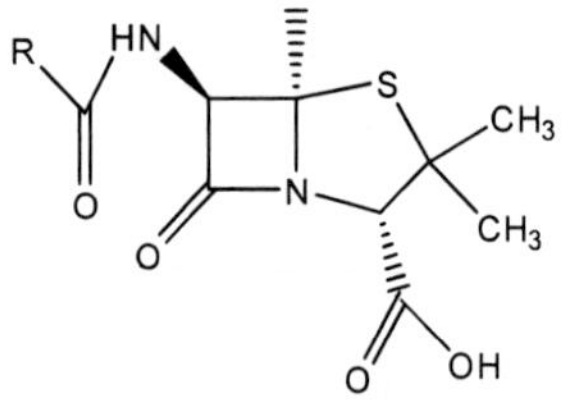

penicillin(11)

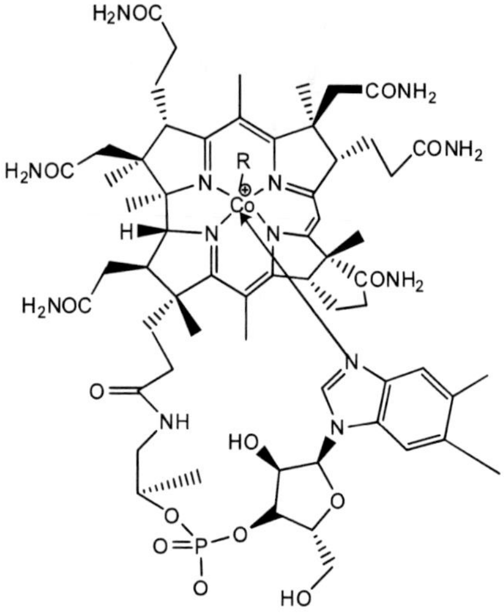

vitamine B12

vitamin B12 (12)

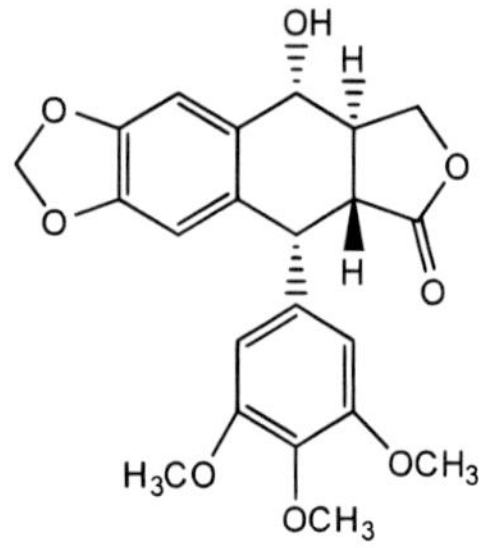

podophyllin (13)

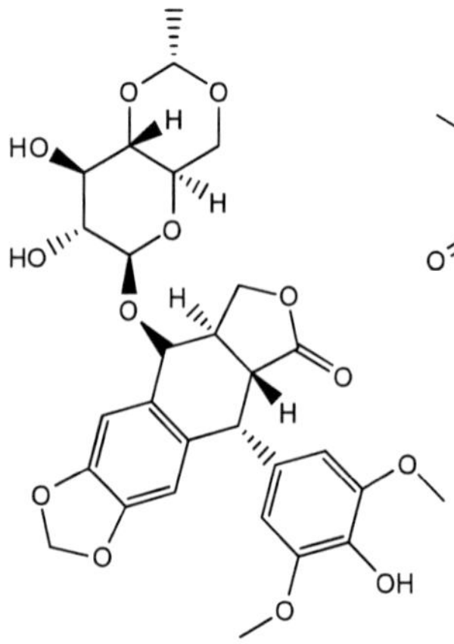

etoposide (14)

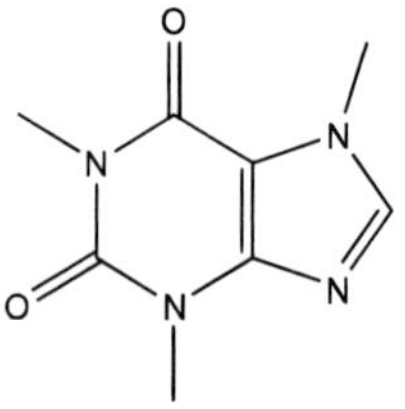

caffeine (15)

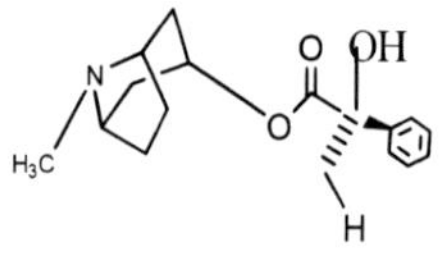

Atropa belladonna(16)

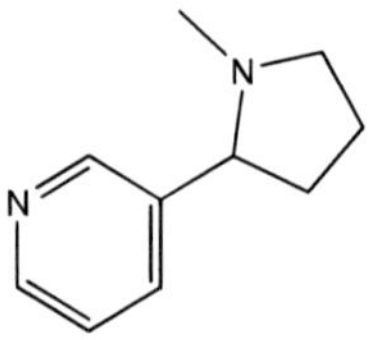

nicotine (17)

Figure 1. (Continued).

Nature-Inspired Phytochemicals ...

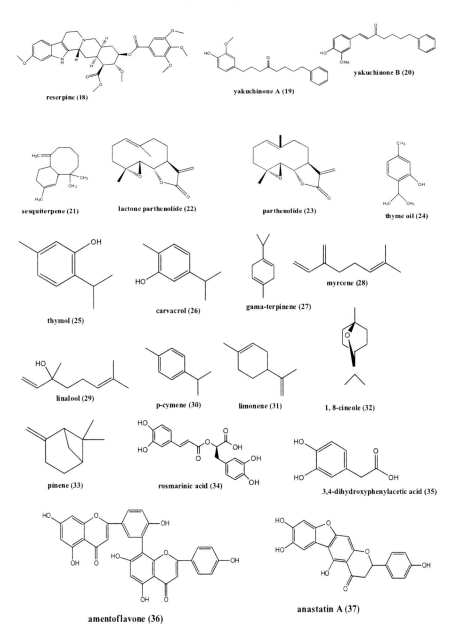

Figure 1. (Continued).

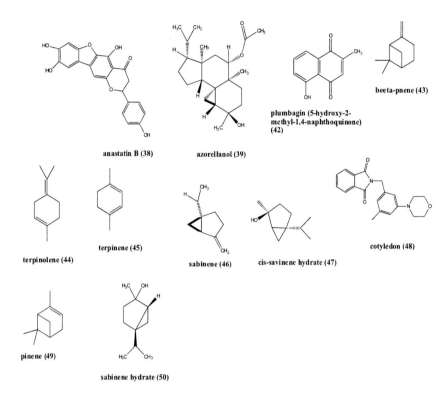

Figure 1. Structures of all compounds (1-50) above mentioned artemisinin (1), taxol (2), camptothecin (3), morphine (4), strychnine (5), quinine (6), codeine (7), digitoxin (8), atropine (9), hyoscine (10), penicillin (11), vitamin B_{12} (12), podophyllin (13), etoposide (14), caffeine (15), Atropa belladonna (16), nicotine (17), reserpine (18), yakuchinone A (1–[4'-hydroxy-3'-methoxyphenyl]-7-phenyl-3-heptanone (19), yakuchinone B (1-[4'-hydroxy-3'-methoxyphenyl]-7-phenylhept-1-en-3-one (20), sesquiterpene (21), lactone parthenolide (22), parthenolide (23), thyme oil (24), thymol (25), carvacrol (26), γ-terpinene (27), myrcene (28), linalool (29), *p*-cymene (30), limonene (31), 1,8-cineole (32), α-pinene (33), rosmarinic acid (34), 3,4-dihydroxyphenyllacetic acid (35), amentoflavone (36), anastatin A (37), anastatin B (38), azorellanol (39), 13-hydroxy-7-oxoazorellane (40), 7-deacetylazorellanol (41), plumbagin (5-hydroxy-2-methyl-1,4-naphthoquinone (42), β-pinene (43), terpinolene (44), terpinene-4-ol (45), sabinene (46), *cis*-savinene hydrate (47), cotyledon (48), pinene (49), and sabinene hydrate (50).

The genus Tanacetum has been used as medicinal plants for over 2000 years. Interest in the genus has been stimulated by the biological activities, particularly as insect anti-feedants, antitumor and antimicrobial agents due to its sesquiterpenoid constituents which are the main constituents of the genus are supposed to be bioactive principles of the plants. Flavonoids and essential oils are also pointed out as active substances in some species. The chemical and biological activities were studied by Goren and co-workers [15].

Results of various projects of Mexican Indian ethnobotany and some of the subsequent pharmacological and phytochemical studies are summarized focusing both on chemical, pharmacological, as well as anthropological (ethnopharmacolgy) aspects. These were reported by Henrich and co-workers [16]. There exists well defined criteria specific for each culture, which lead to the selection of a plant as a medicine. This field research has also formed a basis for studies on bioactive natural products from selected species. The bark of *Guazuma ulmifolia* showed anti-secretory activity (cholera toxin-induced chloride secretion in rabbit distal colon in an Ussing chamber). Active constituents are procyanidins with a polymerisation degree of eight or higher. *Byrsonia crassifolia* yielded proanthocyanidins with (+)-epicatechin units and *Baccharis conferta* showed dose-dependently antispasmodic effect with the effect being particularly strong in flavonoid-rich fractions. Their ethno-pharmacological research led to the identification of sesquiterpene lactones (SLs) like parthenolide as potent and relatively specific inhibitors of the transcription factor NF-kB, an important mediator of the inflammatory process.

Petersen and co-workers [17] proposed rosmarinic acid (**34**) (Figure 1) is an ester of caffeic acid and 3,4-dihydroxyphenyllacetic acid (**35**) (Figure 1). It is commonly found in species of the Boraginaceae and the subfamily Nepetoideae of the Lamiaceae. However, it is also found in species of other higher plant families and in some fern and hornwort species. Rosmarinic acid (**34**) has a number of interesting biological activities, *e.g.*, antiviral, antibacterial, anti-inflammatory and antioxidant. The presence of rosmarinic acid (**34**) in medicinal plant herbs and species has beneficial and health promoting effects. In plants, rosmarinic acid (**34**) is supposed to act as a preformed constitutively accumulated defense compound. The biosynthesis of rosmarinic acid (**34**) starts with the amino acids L-phenylalanine and L-tyrosine.

150 V. Padmavathi, B. Kesava Rao and N. Motohashi

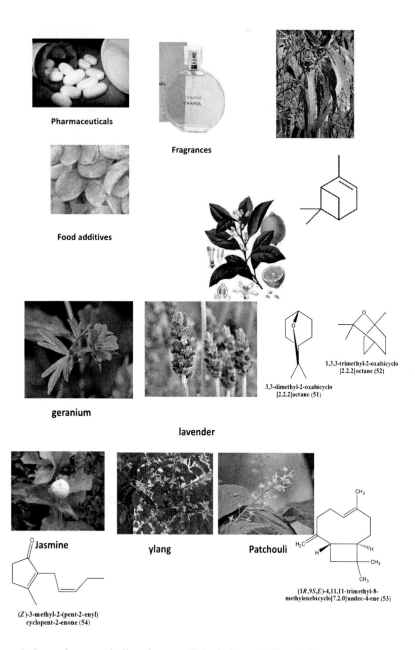

Figure 2. Secondary metabolites from medicinal plants. 3-dimethyl-2-oxabicyclo[2,2,2]-octane (**51**); 1,3,3-trimethyl-2-oxabicyclo[2,2,2]octane (**52**); (*E*,1*R*,9*S*)-4,11,11-trimethyl-8-methylenebicyclo[7,2,0]undec-4-ene (**53**); (*Z*)-3-methyl-2-(pent-2-ynyl)-cyclopent-2-enone (**54**).

Nature-Inspired Phytochemicals ... 151

Some of the secondary metabolites produced by a living organism; which shows distinctive pharmacological effects.

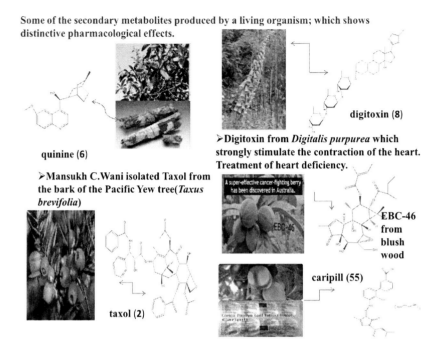

Figure 3. Quinine (**6**), digitoxin (**8**), taxol (**2**), and caripill (**55**) from various medicinal plants.

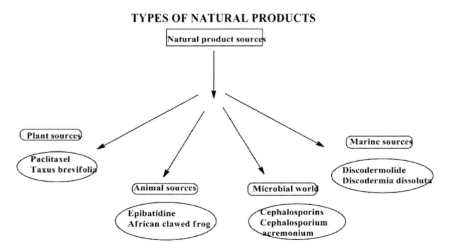

Figure 4. Different types of natural products from natural sources.

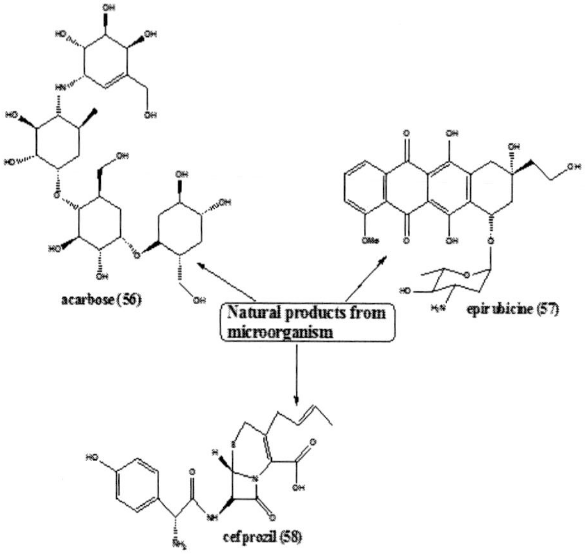

Figure 5. Natural products from microorganism. acarbose (**56**); epirubicin (**57**); cefprozil (**58**).

All eight enzymes involved in the biosynthesis are known and characterized and complementary DNAs (cDNAs) of several of the involved genes have been isolated.

Amentoflavone (**36**) (Figure 1) is found in number of plants with medicinal properties, including *Ginkgo biloba* and *Hypericum perforatum* (St. John's wort). Hanrahan and co-workers [18] developed a rapid and economic semi-synthetic preparation of amentoflavone (**36**) from biflavones isolated from autumnal *Ginkgo biloba* leaves. Several studies have shown that amentoflavone (**36**) binds to benzodiazepine receptors.

New skeletal flavonoids, anastatinA (**37**) and anastatin B (**38**) (Figure 1) were isolated from the methanolic-extract of an Egyptian medicinal herb, the whole plants of *Anastatica hierochuntica*. Their flavonone structures having a benzofuran moiety were determined on the basis of chemical and physiochemical evidences. Anastatin A (**37**) and anastatin B (**38**) were found to show hepatoprotective effects on D-galactosamine-induced cytotoxicity in primary cultured mouse hepatocytes and their

Nature-Inspired Phytochemicals ... 153

activities were stronger than those of related flavonoids and commercial silybin [19].

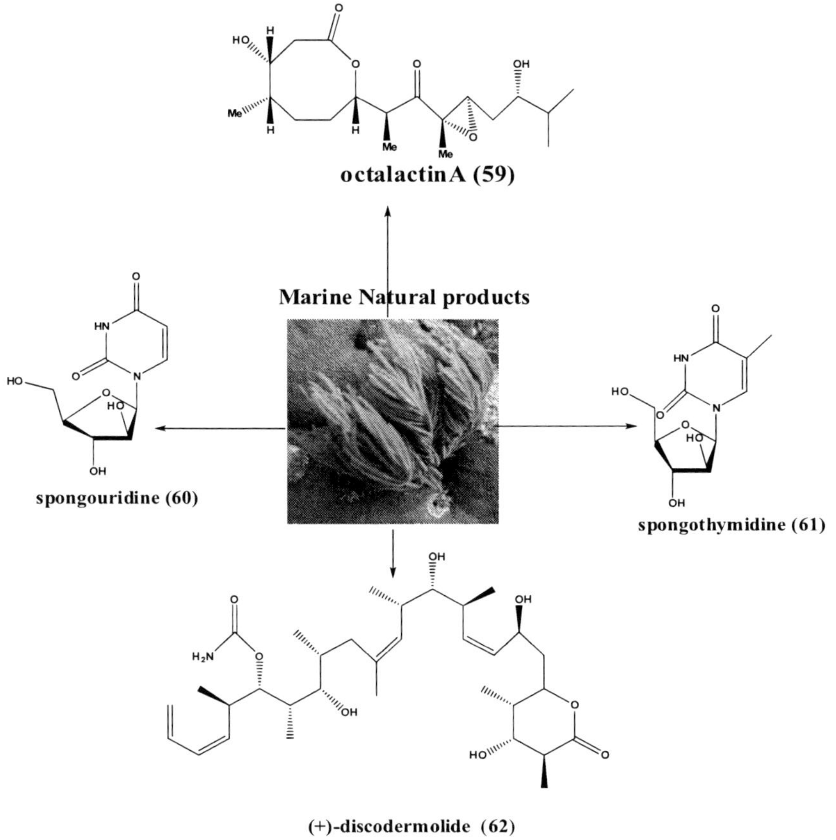

Figure 6. Natural products from marine sources. octalactin A (**59**); spongouridine (**60**); spongothymidine (**61**); (+)-discodermolide (**62**).

Azorella compacta, *Azorella yareta* and Laretiaacaulis (Apiaceae) are native species from the high Andes Mountains, northeastern Chile and they have been traditionally used to treat asthma, colds and bronchitis, illnesses with inflammation and pain as the main symptoms. It was proposed that these medicinal species contain bioactive compounds with anti-inflammatory and analgesic effects. In this context, azorellanol (**39**), 13-hydroxy-7-oxoazorellane (**40**) and 7-deacetylazorellanol (**41**) (Figure 1), three diterpenoids previously isolated only from these plants, were subjected to pharmacological and toxicological evaluation. Their tropical anti-inflammatory and analgesic activities along with acute toxicities or innocuousness were also investigated by Delpofte, and co-workers [20] and results indicate the absence of toxic and side effects in mice.

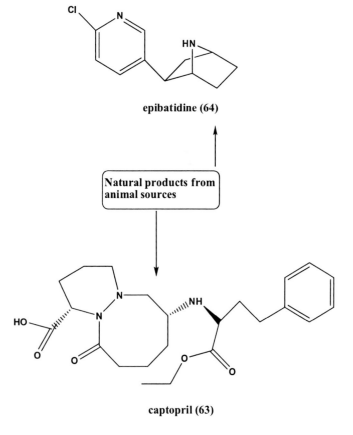

Figure 7. Natural products from animal sources.captopril (**63**); epibatidine (**64**).

All compounds presented dose related inhibition of pain. 13-Hydroxy-7-oxoazorellane (**40**) was the most potent analgesic but it was less effective than sodium -naproxen, the reference drug. Azorellanol (**39**) exhibited the highest topical anti-inflammatory potency on arachidonc acid (AA) and

12-deoxyphorbol 13-tetradecanoate (TPA) induced edema, and its effect was similar to the reference drugs like nimesulide and indomethacin. Probably, its mechanism of action could be explained through the inhibition of cyclooxygenase (COX) activity. Oxidative stress has been implicated in the etiology of a number of human ailments.

Hence, antioxidants especially derived from natural sources are capable of protecting against damage induced by excess reactive oxygen species (ROS) may have potential applications in prevention and/or cure of diseases. Indian medicinal plants provide a rich source of these potentially useful compounds. *Plumbago zeylanica* (known as 'chitrak') and its constituents are credited with potential therapeutic properties including anti-atherogenic, cardiotonic, hepatoprotective and neuroprotective properties. Plumbagin (2-methyl-5-hydroxy, 4 naphthoquinone) **(42)** (Figure 1), isolated from the root of this plant was considered as the active ingredient.

Kumar S and co-workers [21] the structure–function relationships of the naphthoquinone phytochemicals, plumbagin (5-hydroxy-2-methyl-1,4-naphthoquinone) **(42)**, juglone, and menadione, have been studied with regard to antimutagenic and antioxidant activities. Antimutagenicity of these compounds was assessed by the Ames test and RNA polymerase B (rpoB)-based rifampicin resistance assay. Antioxidant potential was evaluated by radical scavenging assays and reducing power measurement. Protection of cells and DNA against gamma radiation-induced oxidative damage was assayed by survival analysis and gel electrophoresis profiling, respectively. On the 1,4-naphthoquinone nucleus, plumbagin **(42)** possesses 5-hydroxyl and 2-methyl functional groups, whereas juglone has only the 5-hydroxyl and menadione only the 2-methyl group. Plumbagin **(42)** showed strong antimutagenic (against ultraviolet and ethyl methanesulfonate) and antioxidant activities, whereas juglone displayed only strong antimutagenic, and menadione only strong antioxidant activities. Thus, these two functional groups (5-OH/2-CH$_3$) play important roles in the differential bioactivity of naphthoquinones. *Escherichia coli*, microarray analysis showed upregulation of the genes rep (replication/repair), ybaK (tRNA editing), speE (spermidine synthesis), and yjfC (glutathionyl spermidine synthesis) by plumbagin (**42**) or juglone, and sodC (superoxide dismutase), xthA (oxidative repair), hycB (electron carrier between hydrogenase 3 and fumarate dehydrogenase), and ligA (formation of phosphodiester bond in DNA) by plumbagin (**42**) or menadione. Studies with *E. coli* single-gene knockouts showed that ybaK and speE, reported to prevent mistranslation, are likely to be involved in the antimutagenicity displayed by juglone, and sodC to be involved in the antioxidant activity of menadione.

The producers of medicinal and aromatic plants need to ensure that their plantation is stocked exclusively with the desired commercial chemical variety. Planting the wrong variety can only prove costly by way of crop removal and replanting operations. The examination of germinant at successive stages of development gave an indication of the chemical changes that takes place between germination and harvest. Using Australian tea tree (*Melaleuca alternifolia* Linn.) as an example, terpenoid biogenetic pathways were found to be initiated at different stages of ontogeny. The cotyledon leaves of common terpinen-4-ol chemical variety seedlings were rich in α-pinene (**33**) (7.4%), β-pinene (**43**) (12.0%) (Figure 1), and terpinolene (**35**) (27.3%). The non-common terpinolene (**35**) variety was found to be rich in 1,8-cineole (**32**) (12.5%) and terpenolene (**36**) (25.4%) and the 1,8-cineole (**32**) variety rich in 1,8-cineole (**32**) (37%) with significant quantities of α-pinene (33) (15.5%), β-pinene (**43**) (23.3%) and terpinolene (**44**) (10.9%) (Figure 1) reported by Southwell and co-workers [22].

Individual leaves of all three chemical varieties were then examined from the emergence of the first true leaves, through to six week old leaf-set-ten material. In the terpinenen-4-ol (**45**) variety, the higher concentrations of terpinolene (**44**), α-pinene (**33**) and β-pinene (**43**) and lower concentrations of terpinen-4-ol (**45**), sabinene (**46**) and *cis*-sabinene hydrate (**47**) (Figure 1) gradually changed, first with the emergence of each new leaf set and then again as each leaf set aged. By the time that leaf-set-ten was 6 week old terpinolene (**44**), α-pinene (**33**) and β-pinene (**43**) levels had fallen, *cis*-savinene hydrate (**47**) had raisen and then fallen and terpine-4-ol (**45**) increased so that all components were now present in concentrations similar to those of mature leaf. In the 1,8-cineole (**32**) and terpinolene (**44**) chemical varieties, some differences were evident, but early leaves better reflected the chemical quality of the mature tree. The cotyledon (**48**) (Figure 1) and early seedling leaf compounds when compared with that of mature leaf from the same chemical variety, was found to be biased toward pinene (**49**) (Figure 1) and terpinolene (**44**) biogenetic pathway constituents. Hence, pinene (**49**) and terpinolene (**44**) biogenesis commences prior to the onset of sabinene hydrate (**50**) (Figure 1) and terpinen-4-ol (**45**) formation. Consequently, early seedling leaf microanalysis is not a good indicator of mature tree quality unless the sequential onset of the biogenetic pathways in tea tree is taken into consideration as reported by Southwell and co-workers [22].

For a long time, plant components have been used in regions such as pharmaceuticals, fragrances, food additives *etc*. Phytochemicals such as 3-dimethyl-2-oxabicyclo[2,2,2]-octane (**51**), 1,3,3-trimethyl-2-oxabicyclo[2,2,2]octane (**52**), ($E,1R,9S$)-4,11,11-trimethyl-8-methylenebicyclo [7,2,0] undec-4-ene (**53**), and (Z)-3-methyl-2-(pent-2-ynyl)-cyclopent-2-enone (**54**) are known as secondary metabolites of medicinal plants (Figure 2). Plant derived-components such as quinine (**6**), digitoxin (**8**), taxol (**2**), and caripill (**55**) have been partially prepared for clinical applications (Figure 3). Therefore, the diverse types of abubdant phyto-source's materials are very important in appearance of novel medicines in drug design (Figure 4). Microorganisms are capable of producing natural products - known as antibiotics - with widely divergent chemical structures such as penicillin (**11**) with β-lactam ring, anthracycline antibiotics with tetracycline ring from *Streptomyces peucetius* var. caesius, acarbose (**56**), epirubicin (**57**) and cefprozil (**58**) (Figure 5). In addition to terrestrial plants, many active compounds - octalactin A (**59**); spongouridine (**60**); spongothymidine (**61**); (+)-discodermolide (**62**) (Figure 6) - have been found from marine organisms. A snake venom captopril (**63**) and epibatidine (**64**) from Anthony's poison arrow frog (Figure 7) are the typical natural products from animal sources.

indole alkaloids (65)

Figure 8. Structure of indole alkaloids (65).

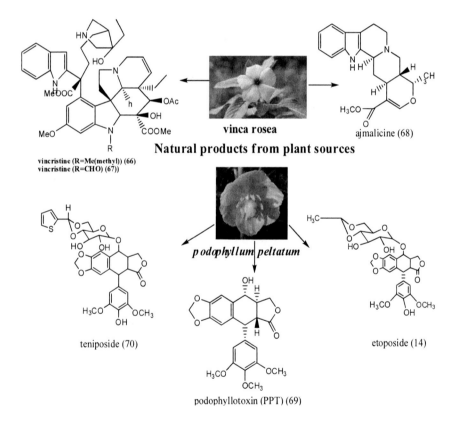

Figure 9. Natural products from plant sources. vincristine (R = Me(methyl)) (66); vinblastine (R = CHO) (67); ajmalicine (68); podophyllotoxin (PPT) (69); etoposide (14); teniposide (70).

1.2. Anti-Cancer Drugs from Plants

Indole alkaloids (**65**) (Figure 8), discovered from the deciduous tree *Camptotheca acuminata*, is also an anticancer agent which has a unique mechanism of action. Indole alkaloids (**65**) and its derivatives are topoisomerase-I (TOP1) inhibitors, and cause cell death by DNA damage. However, indole alkaloids (**65**) itself is too insoluble to be used as a drug but its several water-soluble analogs, namely, topotecan (**76**), and irinotecan (**77**) have been developed as effective drugs [32].

Even though there are a large number of anticancer drugs in market, there is need for the availability of less toxic and more potent anticancer drugs. Most of the cancer drugs even destroy the normal cells, because they are not selective. Natural products act as leads for various synthetically driven anticancer drugs available in the market. The first compound isolated was podophylloforim (**73**), from the plant *Podophyllum peltatum* [23, 24].

Etoposide (**14**) and teniposide (**70**) (Figure 9) are the modified analogs of poelophyllotoin (**72**). These two analogs are used in the treatment of various cancers. It has indole alkaloids which are having anticancer properties. This plant also contains ajmalicine (**68**) (Figure 9) an antihypertensive alkaloid. This plant is used as a remedy for diabetics [25].

An extract of the Pacific yew tree, *Taxus brevifolia* was discovered to possess excellent anticancer properties in 1963, and its active component was isolated only a few years later in 1967 by Monroe Wall and his co-worker, Mansukh Wani [26]. They published their findings as well as the structure of the active component, paclitaxel (taxol) (**2**) [27], in 1971 Susan B. Horwitz, a molecular pharmacologist, established the novel mechanism of action of paclitaxel (**71**) in 1979. Paclitaxel (**71**) irreversibly binds to β-tubulin, thus promoting microtubule stabilization [28]. This tubulin-microtubule equilibrium is essential for cell multiplication, and its stabilization causes programmed cell death [29]. Previously reported anticancer drugs, vincristine (R = Me(methyl) (**66**) (Figure 9), vinblastine (R = CHO) (**67**) (Figure 9) and podophyllotoxin (**69**) (Figure 9) also bind to β-tubulin, but prevent rather than promote microtubule formation.

Paclitaxel (71) was the first compound to be discovered to promote microtubule formation. It has been used in the treatment of several types of cancer, but most commonly for ovarian and breast cancers as well as non-small cell lung tumors [30]. It had sales of $750 million in 2002 and $1.0 billion in 2003 [31]. Shortly after the discovery of paclitaxel and its unique mechanism, several compounds having the same mode of action were discovered. The epothilones (74), discovered from the myxobacterium *Sorangium cellulosum*, possess potential anticancer properties and show high *in vivo* activity, including activity against taxane-resistant cell lines. However, they exhibit moderate *in vitro* cytotoxicity [32, 33]. Several semi-synthetic analogs of epothilones such as dixabepilone (75) have been developed which are currently in Phase II clinical trials for treatment of breast cancer.

Thus medicinal plants and natural products are still ruling the entire medicinal world.

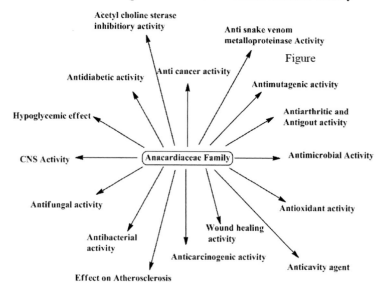

Figure 10. Pharmacological activity of *Anacardiaceae* family.

2. ANACARDIACEAE FAMILY

2.1. Introduction

The *Anacardiaceae* (cashew or *sumac* family) is a family of flowering plants including 83 genera with about 860 species [34], bearing fruits that are drugs and in some cases producing urushiol which is an irritant compound. The *Anacardiaceae* includes numerous genera with several economic importances. *Anacardiaceae* family is subdivided into five tribes as follows: *Anacardieae, Rhoeae, Spondiadeae, Semecarpeae and Dobineaeare* are classified based on morphological characteristics. The numbers of genera and species allocated to these tribes are shown in Figure 10 and Table 1 [35].

2.2. Description of Anacardiaceae Family

Table 1. *Anacardiaceae* family description

Tribe	Affiliated genera
Anacardieae	*Anacardium, Androtium, Bouea, Buchanania, Fegimanra, Gluta (including Melanorrhoea), Mangifera, Swintonia*
Spondiadeae	*Allospondias, Antrocaryon, Choerospondias, Cyrtocarpa, Dracontomelon, Haematostaphis, Haplospondias, Harpephyllum, Koordersiodendron, Lanna, Operculicarya, Pegia, Pleiogynium, Poupartia, Pseudospndias, Sclerocarya, Solenocarpus, Spondias, Tapirira*
Semecarpeae	*Drimycarpus, Holigarna, Melanocytes, Nothopegia, Semecarpus*
Rhoeae	*Abrahamia ined, Actinocheita, Amphipterygium, Apterokarpos, Astronium, Baronia, Blepharocarya, Bonetiella, Campnosperma, Cardenasiodendron, Comocladia, Cotinus, Euroschinus, Faguetia, Haplorhus, Hrrria, Hermogenodendron, Laurophyllus, Lithrea, Loxopterygium, Loxostylis, Malosma, Mauria, Melanococca, Metopium, Micronychia, Mosquitoxylum, Myracrodruon, Ochoterraenea, Orthopterygium, Ozoroa, Pachycormus, Parishia, Pentaspadon, Pistacia, Protorhus, Pseudosmodingium, Rhodosphaera, Rhus, Schinopsis, Schinus, Searsia, Smodinaium, Sorindeia, Thyrsodium, Toxicodendron, Trichoscypha*
Dobineae	*Campylopetalum, Dobinea*

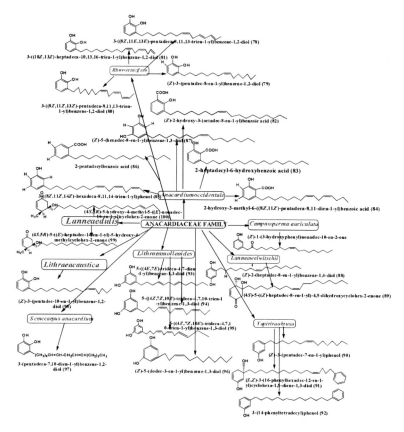

Figure 11. Variousphenolic lipids (**78-100**) isolated from *Anacardiaceae* family. (3-((8Z,11E,13E)-pentadeca-8,11,13-trien-1-yl)benzene-1,2-diol (**78**), (Z)-3-(pentadec-8-en-1-yl)benzene-1,2-diol (**79**), 3-((8Z,11Z,13Z)-pentadeca-8,11,13-trien-1-yl)benzene-1,2-diol (**80**), 3-((10Z,13Z)-heptadeca-10,13,16-trien-1-yl)benzene-1,2-diol (**81**) [*Rhus vernicifera*], (Z)-2-hydroxy-3-(octadec-8-en-1-yl)benzoic acid (**82**), 2-heptadecyl-6-hydroxybenzoic acid (**83**),2-hydroxy-3-methyl-6-((8Z,11Z)-pentadeca-8,11-dienyl)benzoic acid (**84**), 3-((8Z,11Z, 14Z)-hexadeca-8,11,14-trien-1-yl)phenol (**85**), 2-pentadecylbenzoic acid (**86**), (Z)-5-(hexadec-8-en-1-yl)benzene-1,3-diol (**87**) [*Anacardium occidentale*], (Z)-2-(heptadec-8-en-1-yl)benzene-1,4-diol (**88**) [*Campnosperma auriculata*], (4S)-5-((Z)-heptadec-8-en-1-yl)-4,5-dihydroxycyclohex-2-enone (**89**), (Z)-3-(heptadec-7-en-1-yl)phenol (**90**), (S,Z)-3-(16-phenylhexadec-12-en-1-yl)cyclohexa-1,5-diene-1,3-diol (**91**), 3-(14-phenyltetradecyl) phenol (**92**) [*Tapiri raobtusa*], 5-((4E,7E)-trideca-4,7-dien-1-yl)benzene-1,3-diol (**93**), 5-((4Z,7Z,10Z)-trideca-4,7,10-trien-1-yl)benzene-1,3-diol (**94**), 5-((4Z,7Z,10E)-trideca-4,7,10-trien-1-yl)benzene-1,3-diol (**95**), (Z)-5-(dodec-3-en-1-yl)benzene-1,3-diol (**96**) [*Lithraeamolleoides*], 3-(pentadeca-7,10-dien-1-l)benzene-1,2-diol (**97**) [*Semecarpus anacardium*], (Z)-3-(pentadec-10-en-1-yl)benzene-1,2-diol (**98**) [*Lithrae acaustica*], (4S,5R)-5-((E)-heptadec-14-en-1-yl)-5-hydroxy-4-methylcyclohex-2-enone (**99**), (4S,5R)-5-hydroxy-4 methyl-5-((E)-nonadec-16-en-1-yl)cyclohex-2-enone (**100**) [*Lanne aedulis*].

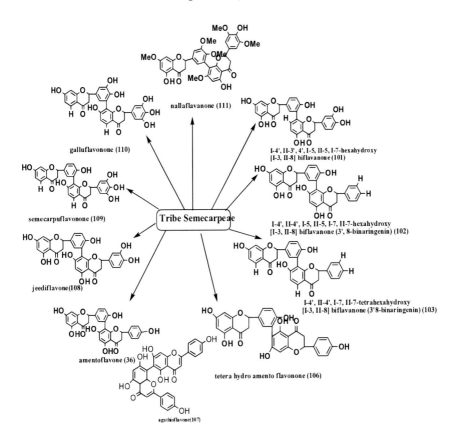

Figure 12. Eleven flavonoidsisolated from *Anacardiacea* family of tribe Semecarpeae. amentoflavone (**36**), I-4^1,II-3^1,4^1,I-5,II5,I7-hexahydroxy[I3,II8]biflavonone (**101**), I-4^1,II-4^1,I-5,II-5,I-7,II-7-hexahydroxy[I-3,II-8]biflavonone (3^1,8-binaringenin) (**102**), I-4^1,II-4^1,I-7,II-7-tetrahexahydroxy [I-3,II-8]biflavonone (3^1,8-billiquiritigenin) (**103**), (*E*)-3-(pentadec-8-enyl)benzene-1,2-diol (**104**) (Figure 21), 3-((7*E*,10*E*)-pentaca-7,10-dienyl)benzene-1,2-diol (**105**) (Table 2), tetrahydroamentoflavonone (**106**), agathisflavone (**107**), jeediflavone (**108**), semecarpuflavonone (**109**), galluflavonone (**110**) and nallaflavanone (**111**).

2.3. Phytochemistry of Anacardiaceae

Phenolic lipids were isolated from *Anacardium occidentale, Lannea edulis, Campnosperma auriculata, Lannea welwitschii, Lithraea caustica, Gluta renghas, Lithaea molleoides, Lannea edulis, Melanorrhoea usitate,*

Mtopium brownie, Pistachia vera, Rhus semeliata, Rhus succedanea, Rhus vernicifera, Schinopsis brasiliensis, Semecarpusgardneri, S. obscura, S.petaltae, S. walkeri, Semecarpus anacardium, Spondias mombin, Tapirira guianensis, Tapirira obtuse (Figures 11-15).

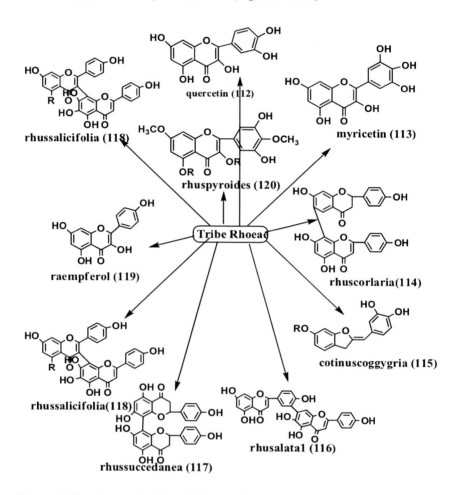

Figure 13. Ten flavonoids (112-120) isolated from Anacardiacea family of Spondiadeae tribe Rhoeae [62, 63]. quercetin(e) (112), miricetina (myricetin. 113), rhuscorlaria (114), cotinuscoggygria (115), rhusalatal (116), rhussuccedanea (117), rhussalicifolia (118), kaempferol (119), rhussalicifolia (118), rhuspyroides (120).

Nature-Inspired Phytochemicals ...   165

Figure 14. Flavonoids isolated from *Anacardiacea* family of tribe Sapondiadeae [62]. Kaempferol (**119**), kaempferide (**121**), 8,8-dimethyl-10-(3-methylbut-2-en-1-yl)-2-phenyl-2,3,7,8-tetrahydropyrano[3,2-g]chromen-4(6*H*)-one (**122**).

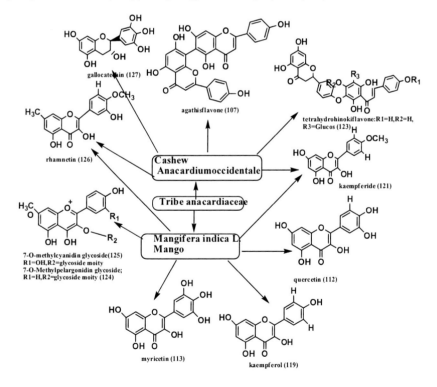

Figure 15. Some of the chemical constituents recently isolated from various tribes. agathisflavone (**107**), quercetin (**112**), miricetina (myricetin. **113**), kaempferol (**119**), kaempferide (**121**), tetrahydrohinokiflavone (**123**), 7-*O*-methylpelargonidin glycoside (**124**), 7-*O*-methylcyanidin glycoside (**125**), rhamnetin (**126**), gallocatechin (**127**).

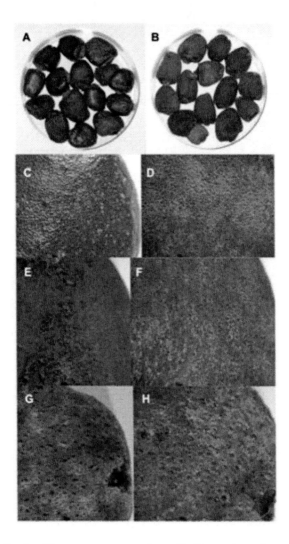

Figure 16. A: *Seeds of Semecarpus anacardium*.; B: Change in surface color of *S. anacardium* seeds due to elimination of the dark-brown pigments after concentrated sulfuric acid treatment.; C: Surface of the seed with granular undulation before sulfuric acid treatment.; D: Absence of granular undulation and presence of innumerable pores on the surface of the seed after concentrated sulfuric acid treatment for 20 min.; E: Surface of seed treated with sulfuric acid for 10 min. Presence of residual pigment indicates incomplete removal of the pigment.; F: Surface of seed after 20 min sulfuric acid treatment. Absence of pigment and presence of numerous small pores suggests complete removal of the pigment layer; G and H: Surface of the seeds after 30 and 40 min of sulfuric acid treatment. Appearance of larger pores indicates erosion of the seed coat.

3. *SEMECARPUS ANACARDIUM* L.F.

3.1. Introduction

In the 21st century herbal drugs are contributing much to human health. In our tradition, especially, natural medicine improves the inner immune system of the human body and no adverse effects could be observed. Hence, the herbal drug acts more effectively than the modern medicine. In our survey, we found that, the *Anacardiaceae* family has 83 genera and 860 species existing as trees, shrubs and vines. Out of these, the commonly known Cashew family or Sumac family has tremendous commercial importance. Due to the presence of several anticancer drugs isolated from *Anacardiaceae* family, we have selected *Semecarpus anacardium* L.f. (Photo is available- See Figure 18) [63-65], for its high medicinal value in Ayurveda and Siddha systems and isolated several active constituents.

The present study reveals the most interesting innovations done on the isolation of phytochemical constituents from *Semecarpus anacardium* (bhallataka) and synthesis of bhallataka biodiesel, biolubricants, novel applications and their biological activities.

3.2. General Profile of *Semecarpus anacardium* L.f.

3.2.1. Taxonomical Classification

- Kingdom: *Plantae*
- Subkingdom: *Tracheobionta*
- Superdivision: *Spermatophyta*
- Division: *Magnoliophyta*
- Class: *Magnoliopsida*
- Subclass: *Rosidae*
- Order: *Sapindales*
- Family: *Anacardiaceae*
- Genus: *Semecarpus*

- Species: *anacardium*
- Botanical name: *Semecarpus anacardium* L.f. [66].

3.2.2. Synonyms

- *Anacardium latifolium* Lam. (Anacardiaceae), *Anacardium officinarum* Gaertn. (Anacardiaceae)
- *Anacardium orientale* (Anacardium): *Semecarpus latifoliapers*
- *Telugu*: Nallajeedi, Simidijeedi, Chimidi, Chakalijeedi, Seedi, Maraka, bhilwa, bhilwan, geru, bhela.
- *English:* marking nut tree, marsh nut, Oriental cashew nut, phobinuttree, varnishtree
- *Hindi:* Bhela (bhel), Bhelwa, Bhawa, Bhilwa
- *Tamil:* Erimugi (Erimuki)
- *Gujarati:* Bhilamu
- *Russian:* Semekarpusanakardii
- *Marathi:* Bibba, Bibha.
- *Punjabi:* Bhilawa.
- *Oriya:* Bhollataki
- *Kannada:* Bhallataka, Bhallika, Goddugeru, Karigeri.
- *Greek:* Semekarposanakardion, Semeokarpos to anacardion.
- *Italian:* Anacardio-orientale.
- *Sanskrit:* Antahsattva, Arusharah, Aruskara (Arukara), Arzohita,
- Ballata (Bhallata, Ballata), Bhallataka (Bhalltaka), Bhallatakah, Viravrksa.
- *Assamese:* Bhelaguti.
- *Bengali:* Bhela (bhela), Bhelatuki.
- *Kannada:* Bhallataka, Bhallika, Goddugeru, Karigeri.
- *Malayalam:* Alakkuceru (alakkuceru), Chera.
- *Marathi:* Bibba, Bibha.
- *Oriya:* Bhollataki. Punjabi: Bhilawa.
- *Russian:* Semekarpus Anakardii
- *Danish:* Ostindiskelefantlus.

- *Dutch:* Malakkanoot, Oostindische, acajounoot, Oost-Indische olifantsluis.
- *Arabic:* Habb al fahm, Habb al qalb.
- *French:* Anacarde d'Orient, Noix À Marquer, Noix Des Marais.
- *German:* Anakardien-Herznub, Malakkanub, Elefantenlausbaum, Ostindische Elefantenlaus
- *Japanese*: Anacardium orientale
- *Nepaese:* Kaagbhalaayo
- *Portuguese:* Anacárdio oriental
- *Spanish:* Anacardio oriental

3.2.3. Ayurvedic Properties

Rasa - Katu, Tikta, Kashaya, Madhuraguna - Laghu, Snigdha, Teekshnaveerya (sharp/quick potency) - Ushnavipaka – Madhuradoshaghnata – Kaphavatashamaka Rogaghnata - Indralupta, Mastishkadaurbalya, Nadidaurbalya, Apasmara, Gridhrasi, Urustambha, Amavata, Vatavyadhi, Pakshaghata, Agnimandya, Vibandha, Anaha, Gulma, Udara, Pleehodara, Grahani, Arsha, Krimi, Hriddaurbalya, Granthishotha, Kasa, Shwasa, Prameha, Shukradaurbalya, Dhwajabhanga, Kashtartava, Kushtha, Shwitra, Vatarakta, Vrana, Jwara, Daurbalya, Pleehavriddhi Karma - Pittavardhaka, Sphotajanaka, Sheetaprashamana, Vishaghna, Medhya, Nadibalya, Deepana, Pachana, Bhedann, Yakriduttejaka, Krimighna, Hridayottejaka, Shothahara, Kaphanissaraka, Vrishya, Kamottejaka, Shukrasravavridhikara, Vajikarana, Garbhashayottejaka, Swedajanana, Kushthaghna, Jwaraghna, Rasayana, Chhedana doses - Oil - 10-20 drops; Fruit - 1-2 gram in Kshirapaka form [67].

3.2.4. Siddha Properties

Siddha Name: Chenkottai, Vallaathi,Erimugi, Nandhividai, Poothanasanam, Sombalam, Virasagi, Tembarai, KalagamSuvai (Taste): - Kaippu (Bitter) Veeriyam (Potency): - Veppam (Hot) Vipakam (Transformation): - Karppu (Pungent), Ceikai (Pharmacological action): - Udalthetri (Alterative), Punnakki

(Caustic) Gunam (Uses): - Leprosy, Tuberculosis, Leucoderma, Piles, Vatha diseases and Gastric ulcer. Siddha pharmaceutical preparations: - Serankottainei, RasaganthiMezhugu, Idivallathi, Mahavallathiilagam, Nandi Mezhugu, Neerandasanjeevithylum.

3.2.5. Morphology

Moderate to large sized deciduous tree [68], attaining height of 15m - 25m, with large stiff leaves.

- *Leaves:* crowded at the ends of branches, alternate 30-60 cm long and 12-30 cm broad, obovate /oblong with prominent secondary leaves.
- The tree is leafless in March- April.
- *Bark:* 2-5 cm thick, dusky gray, blackish, with irregular quadrangular plates separated by narrow longitudinal furrows.
- *Flowering:* Occurs in May-June. Flowers: dioceous, small up to 0.8 cm, dull greenish yellow in terminal panicles, sepals and petals 5 each. Stamens: 10, Ovary: unilocular. Ovule: 1. pendulous; style 3, free.
- *Fruits:* (drupe) obliquely ovoid more than1 cm girth and 2.5 cm length fleshy hydrocarp with attached nut/seed. Kernel present inside hard shell is edible but sometimes causes cutaneous eruptions.
- *Seeds:* Seeds are collected between December to March. Average seed collection will be 460-880 seeds/kg. Orange colored receptacle turns black after drying. Collection is done by manual means.
- *Soil:* No specific soil affinity.
- *Distribution:* Dry deciduous forests of all districts.
- *Germination:* Bhuban Mohan Pandal et al. [69] proved that to overcome the seed viability, soaking seeds in concentrated sulfuric acid (H_2SO_4) helped in eliminating the pigmented layer from the surface of the seeds thereby making way for the phenolics to be

released from seeds in washings which followed the H_2SO_4 treatment. Thus leaching of phenolics by the seeds in culture medium was reduced by treating them with H_2SO_4 prior to surface sterilization with sodium hypochlorite (NaOCl). Changes in the seed surface following acid treatments for varying periods (10, 20, 30, 40 min) were studied microscopically. Concentrated H_2SO_4 treatment for 20 min helped to increase the seed coat permeability and excretion of phenolics from the seeds. It also acted as a surface sterilant to a limited extent. Frequency of germination was increased to 63% when seeds were treated with H_2SO_4 for 20 min while exposure for 10 min was ineffective in controlling contamination, whereas longer exposures (30 and 40 min) injured the embryos. Elimination of sucrose in the medium, improved the germination from 50 to 63%. A gradual loss of seed viability from 33 to 4% following storage for 5 months was demonstrated *in vitro* (Figures 16, 17, 18).

3.2.6. Traditional Uses

The gum, fruit and oil of *Semecarpus anacardium* L.f. are used medicinally [68].

The brown gum that exudes from the stem is considered to be useful for the cure of venereal, scrofulous and leprous diseases and also used for preparing varnish.

The ripe fruit is an acrid stimulant, digestive, sedative, antispasmodic, alternative, nervine, tonic and escharotics for internal use.

The fruit is first treated under the flame of cow dung and washed with water. The fruit thus prepared is given with butter and administered to the patients suffering from dyspepsia, piles, skin diseases, nervous debility, worms, palsy and epilepsy. In small doses, it is used as an alternative in scrofula, venereal diseases and leprosy for the treatment of piles.

The fruit is given generally in the form of confection in some places. In some places, the nut is given internally for the treatment of asthma and worms after being first soaked in buttermilk for some time.

The juice is a powerful escortic, internally, it is given in small doses of 1-2 minims mixed with some bland oil or butter or oil seeds for the treatment of leprous and scrofulous infections, syphilis, all kinds of skin diseases, palsy, epilepsy, nervous debility, neuralgia and other diseases of the nervous system, asthma, dyspepsia, piles.

The oil that is extracted from the nut is a powerful escortic antiseptic and cholagogue. A single nut is heated over a flame and the oil that exudes is collected in a vessel containing a pint and half of milk. This mixture of oil and milk is taken everyday for the treatment of cough and relaxation of the uvula and the palate.

Externally, the bruised nut is used as an abortifacient by placing it on the mouth of the uterus.

For the cure of piles, the vapor from burning shells is allowed to apply on the painful swollen parts.

Figure 17. A: Horizontally cut seeds showing, epicarp (Pc), mesocarp (Mc), endocarp (Ec); B: Horizontally cut acid treated (20 min) seed showing, eroded epicarp (Pc), mesocarp (Mc) and endocarp (Ec) are unaffected; C: Seedsection cut vertically. Intact micropylar region before acid treatment is apparent; D: Surface appearance of micropylar (Mp) region of intact seed; E: Section of acid-treated (20 min) seed cut through Mp region vertically. The Mp region appears distorted but no cracks were seen; F: Mp region of seed after acid treatment for 40 min. Appearance of crack suggests damage in this region due to over exposure in sulphuric acid; G: *In vitro*-raised seedlings *Semecarpus anacardium* obtained from seeds treated with concentrated sulphuric acid for 20 min before surface sterilization.

Figure 18. Images of *Semecarpus anacardium* L.f.

The juice from the shell and the oil extracted from the nut either by expression or with the aid of heat, are powerful counterirritants and vesicants. They are used locally for rheumatism, sprains, leprotic nodules, piles, bruises and inflammations of bones and joints.

The stem, bark, leaves, roots and fruit are used for the treatment of snake bite.

The nut yields a powerful and bitter substance used everywhere in India as a substitute for marking ink for clothes by washer men, hence it is frequently called dhobi nut. It gives a black color to cotton fabrics, but before application it must be mixed with limewater as a fixator.

The fruits are also used as a dye. They are also used in Ayurveda medicine as astringent, heat generation appetizer, digestant, rejuvenative, aphrodisiac herb and alleviate the skin and rheumatic disorders, nervous debility, leprosy, asthma, neuralgia, psoriasis, tumors, epilepsy, worm infestation, abdominal disorders, polyuria, leucoderma. The fleshy cups, on which the nuts rest and the kernels of the nuts are eaten (cures diabetes).

Folk medicine *Semecarpus anacardium* is a one of most popular medicinally valuable plant in world of Ayurveda medicine. Charak, Sushrut and Vagbhatt - the main three treatises of Ayurveda medicine - have described the medicinal properties of *Semecarpus anacardium* and its formulation. Bhallataka (*Semecarpus anacardium*) is used both, internally as well as externally.

Detoxified nut of *Semecarpus anacardium* used in Ayurveda medicine for skin diseases, tumors, malignant growths, fevers, haemoptysis, excessive menstruation, vaginal discharge, deficient lactation, constipations, and intestinal parasites (Charaka a Sanskrit text on Ayurveda (Indian traditional medicine), and Sushruta (an ancient Sanskrit text on medicine and surgery)), before using *Semecarpus anacardium* for medicinal purpose, it's necessary to detoxifying it because it is highly toxic for body if not use properly. Number of detoxification methods have been recorded the most common detoxification method involves rubbing of *Semecarpus anacardium* seeds with brick powder and then washing the seeds with warm water. The second common recommended method is to tie the seeds in muslin cloth and suspended it in vessel containing coconut water, then heated for about 3 hrs continuously [70]. The seeds oil is mainly used for medicinal purpose. Seeds are generally boiled in milk and the milk is consumed.

The seeds oil is used in minimum possible quantity, typically mixed with food items or mustard oil. Externally, the oil is applied on wounds to prevent pus formation and better healing of wounds. It works well, when medicated with garlic, onion and ajavayana in sesame oil. In glandular swellings and filariasis, the application of its oil facilitates to drain out the discharges of pus and fluids and eases the conditions.

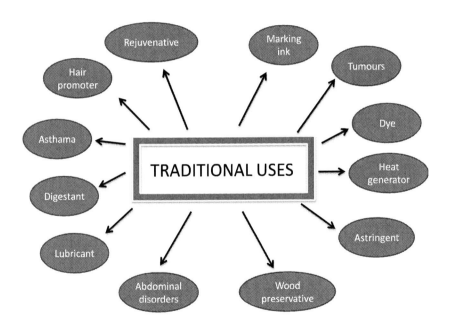

Figure 19. Traditional uses of *Semecarpus anacardium* L.f.

It is also used as a brain tonic, blood purifier and haematinic tonic. The combination, *Semecarpus anacardium*, *Terminalia chebula*, *Sesamum indicum* L. seeds powders with jaggery, has excellent results in chronic rheumatic disorders. In dysmenorrheal (painful menstruation) and oligomenorrhea (scanty menstruation), the medicated milk or its oil is salubrious. It reduces the urinary output, hence beneficial in diabetes of kapha-type; Bhallataka is the best rejuvenative (rasayana) for skin ailments, vata disorders and as a preventive measure to increase the body resistance. Winter is the best season for its usage (Figure 19).

3.3. Compounds and Pharmacological Effects in *Semecarpus anacardium* L.f.

3.3.1. Phytochemical Profile

Phytochemical examination revealed 3.85% of total ash, 0.33% of acid (Figure 20) (Table 2).

Table 2. Compounds isolated from nuts of *Semecarpus anacardium* L.f.

S. No	Compound name	Molecular formula	Structure	Molecular weight	Ref.
1	bhilwanol (**128**)	$C_{21}H_{32}O_2$		316.48	[71]
2	anacardic acid (**129**)	$C_{22}H_{36}O_3$		348.52	[72]
3	1-pentadeca-7,10-dienyl-1,3-dihydroxybenzene (**130**)	$C_{21}H_{38}O_2$		322.53	[73]

S. No	Compound name	Molecular formula	Structure	Molecular weight	Ref.
4	1-pentadeca-8-enyl-2,3-dihydroxy benzene (**131**)	$C_{15}H_{22}O_2$		234.33	[73]
5	I-4¹,II-3¹,4¹,I-5,II5,I7-hexahydroxy [I3,II8]biflavonone (**101**)	$C_{30}H_{22}O_{10}$		542.49	[74]
6	I-4¹,II-4¹,I-5, II-5,I-7,II-7-hexahydroxy [I-3, II-8]biflavonone (3¹,8-binaringenin) (**102**)	$C_{30}H_{22}O_9$		526.49	[74]
7	I-4¹,II-4¹,I-7,II-7-tetrahexahydroxy [I-3,II-8] biflavonone (3¹,8-billiquiritigenin) (**103**)	$C_{30}H_{22}O_7$		494.49	[74]
8	(E)-3-(pentadec-8-enyl)benzene-1,2-diol (**104**)	$C_{21}H_{34}O_2$		318.26	[75]

Table 2. (Continued)

S. No	Compound name	Molecular formula	Structure	Molecular weight	Ref.
9	3-((7E,10E)-pentaca-7,10-dienyl)benzene-1,2-diol (105)	$C_{21}H_{32}O_2$		316.48	[75]
10	Tetrahydro-robustaflavone (132)	$C_{30}H_{22}O_{10}$		542.49	[76]
11	Tetrahydro-amentoflavanone (106)	$C_{30}H_{22}O_{10}$		542.49	[76]
12	amentoflavone (36)	$C_{30}H_{18}O_{10}$		538.46	[76]
13	jeediflavone (108)	$C_{30}H_{22}O_{11}$		558.12	[77, 78]
14	Semecarpuflavanone (109)	$C_{30}H_{22}O_{10}$		542.12	[79]

S. No	Compound name	Molecular formula	Structure	Molecular weight	Ref.
15	galluflavanone (110)	$C_{30}H_{22}O_{11}$		558.12	[80]
16	nallaflavanone (111)	$C_{36}H_{34}O_{13}$		674.65	[81]
17	semecarpetin (133)	$C_{34}H_{30}O_{10}$		598.18	[82]
18	anacarduflavonone (85)	$C_{34}H_{34}O_{12}$		634.21	[83]
19	anacardoside (135)	$C_{19}H_{28}O_{12}$		448.42	[84]

Table 2. (Continued)

S. No	Compound name	Molecular formula	Structure	Molecular weight	Ref.
20	3-pentadecyl benzene (136)	$C_{21}H_{39}O_2$		320.51	[85]
21	(Z)-3-(pentadec-8-enyl)benz-1,2-diol (137)	$C_{21}H_{34}O_2$		318.48	[85]
22	3-((8Z,11Z)-pentaceca-8,11-dienyl) benzene-1,2-diol (138)	$C_{21}H_{32}O_2$		316.48	[85]
23	3,4,2¹4¹-tetrahydroxychalcone (139)	$C_{15}H_{12}O_5$		272.25	[86]
24	7,3¹,4¹-trihydroxyflavone (140)	$C_{16}H_{12}O_4$		268.07	[86]

Nature-Inspired Phytochemicals ... 181

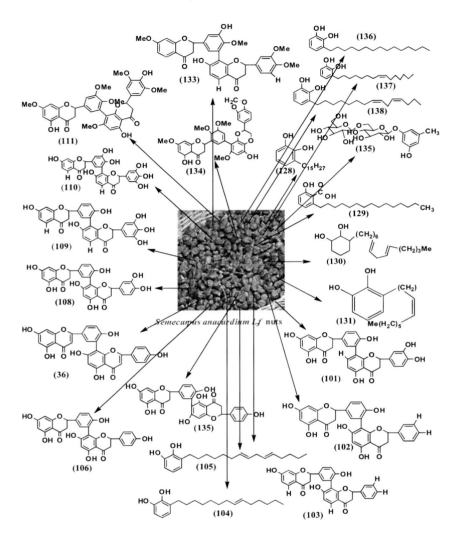

Figure 20. Phytochemical constituents from *Semecarpus anacardium* L.f. amentoflavone (**36**), I-4¹,II-3¹,4¹,I-5,II5,I7-hexahydroxy[I3,II8]biflavonone (**101**), I-4¹,II-4¹,I-5,II-5,I-7,II-7-hexahydroxy[I-3,II-8]biflavonone(3¹,8-binaringenin) (**102**), I-4¹,II-4¹,I-7,II-7-tetrahexahydroxy[I-3,II-8]biflavonone(3¹,8-billiquiritigenin) (**103**), (*E*)-3-(pentadec-8-enyl)benzene-1,2-diol (**104**), 3-((7E,10E)-pentaca-7,10-dienyl)benzene-1,2-diol (**105**), tetrahydroamento flavone (**106**), jeediflavanone (**108**), semecarpuflavanone (**109**), galluflavanone (**110**), nallaflavanone (**111**), bhilwanol (**128**), anacardic acid (**129**), 1-pentadeca-7,10-dienyl-1,3-dihydroxybenzene (**130**), 1-pentadeca-8-enyl-2,3-dihydroxybenzene (**131**), tetrahydrorobustaflavone (**132**) (Figure 22), semecarpetin (**133**), anacardoflavonone (**134**), anacardoside (**135**), 3-pentadecyl benzene (**136**), (Z)-3-(pentadec-8-enyl)benz-1,2-diol (**137**), 3-((8Z,11Z)-pentaceca-8,11-dienyl) benzene-1,2-diol (**138**).

3.3.1.1. Compounds in Nuts of Semecarpus anacardium L.f.

Naidu D.S [70] extracted and prepared the juice from the pericarp of the fruit which was found to contain catechol fixed oil and anacardol ($C_{18}H_{13}O_3COOH$), in which the astringent properties of the juice are due to two phenolic acids.

Based on the investigation of Siddiqui [71], pericarp of the fruit contains a bitter and powerful astringent.

A monohydroxy phenol of 0.1% was contained in the extract this was named as semecarpol (bp 185 to 190°C) that congeals below 25°C to a fatty mass.

Proportion of O-dihydroxy compound was 46% in the extract (15% of the nuts). This has been called bhilawanol (**128**) (Figure 20) (This distills at 225 to 226°C and congeals below 5°C).

Tarry, non volatile, dark brown corrosive residue is forming above 50°C in large retorts. It is a dark, vesicant 18% of the nuts.

It has been further showed that the pericarp contains 20% of oil which can be distilled out from nuts by slow heating to 350°C in large retorts. It is a dark, viscous, highly vesicant liquid which contains bhilawanol (**128**) and other compounds.

Pillay and Siddique [71] isolated a phenolic liquid which was named as bhilawanol (**128**) and regarded it as a catechol derivative with $C_{15}H_{27}$ chain. Many of the properties of marking nut oils could be easily explained by the catechol and lipid soluble C_{15} chain. During the exposure to air, the catechol ring might be oxidized to an *ortho*-quinone which might impart the dark color and subsequently polymerization takes place. The vesicant nature and the inedible pigmentation may be closely connected with the rapid formation of the *ortho*-quinone intermediate. The absorption of the oil by the skin is obviously due to the lipid soluble nature of bhilawanol (**128**).

Later, Chattopadhya and Khare RL [72] isolated anacardic acid (**129**) (Figure 20) from the oil of the nuts. The isolated acid and its sodium salt are found to be potent towards antihelminthic and their acivity was compared with that of piperazine.

Later, Govindachari TR *et al.* [73] revised the Rao and Row work.

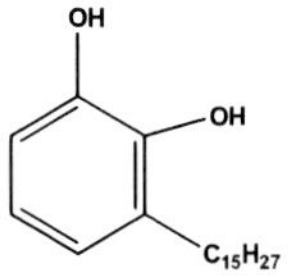

bhilawanol (**128**)

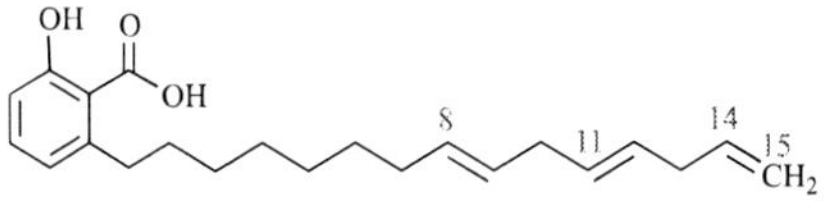

anacardic acid (C$_{22}$H$_{36}$O$_3$. **129**)

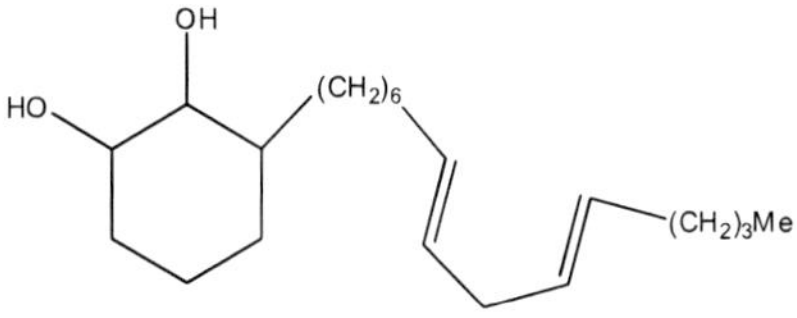

1-pentadeca-7,10-dienyl-1,3-dihydroxybenzene (**130**)

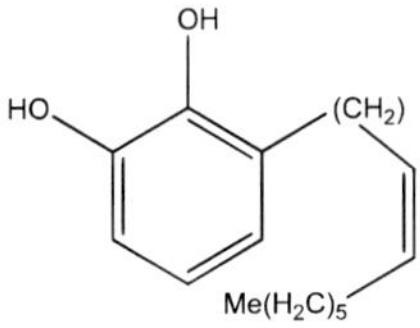

1-pentadeca-8-enyl-2,3-dihydroxybenzene (**131**)

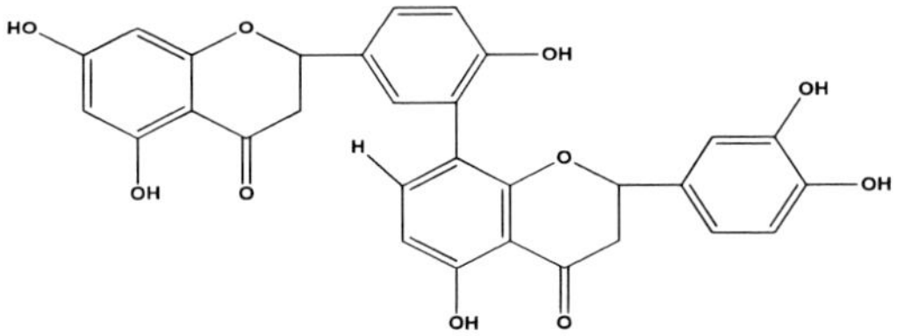

I-4^1,II-3^1,4^1,I-5,II-5,I-7-hexahydroxy[I-3, II-8]biflavonone (**101**)

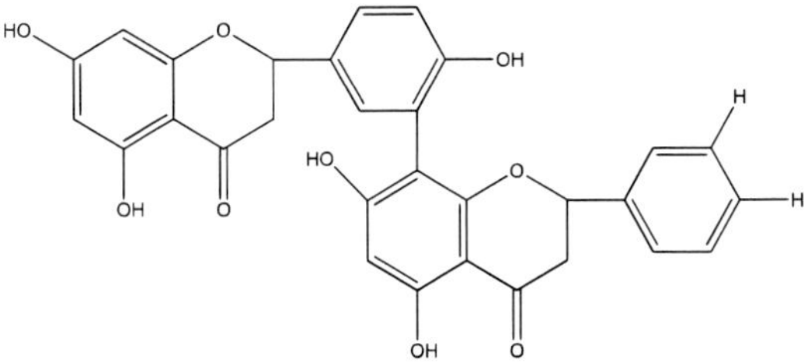

I-4^1,I-4^1,I-5,II-5,I-7,II-7-hexahydroxy[I-3,II-8]biflavonone (3^1,8-binaringenin) (**102**)

Figure 21. (Continued).

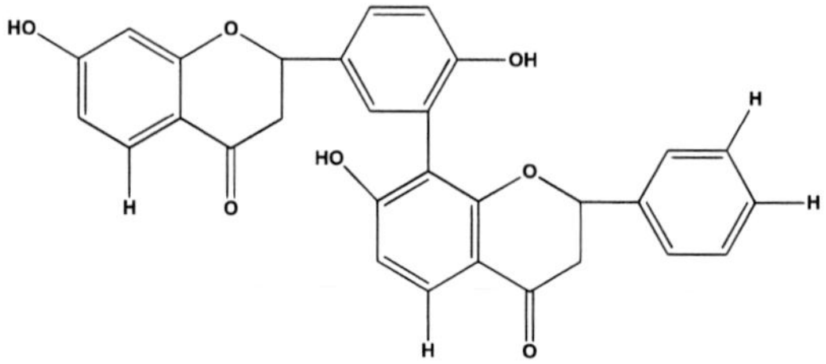

I-4¹,II-4¹,I-7,II-7-tetrahexahydroxy[I-3,II-8]biflavonone
(3¹,8-binaringenin) (**103**)

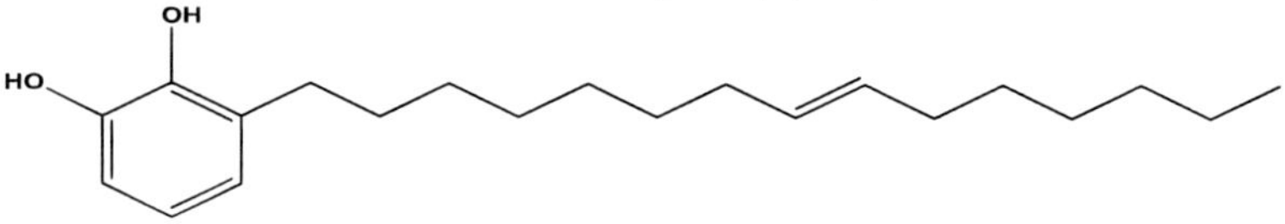

E(-3-(pentadec-8-enyl)benzene-1,2-diol (**104**)

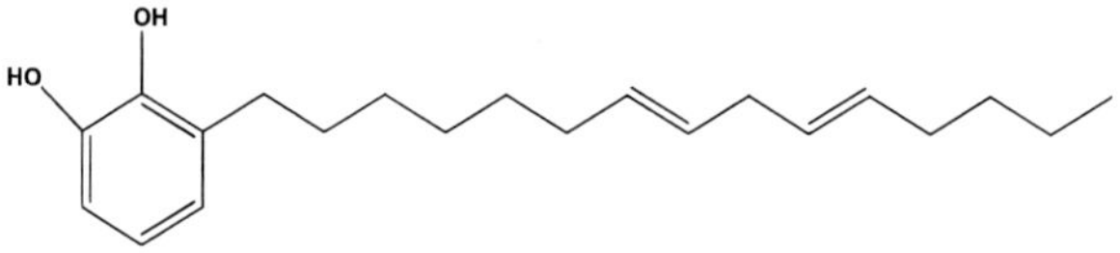

3-((7*E*, 10*E*)-pentadeca-7,10-dien-1-yl)benzene-1,2-diol (**105**)

Figure 21. Bhilawanol (**128**), anacardic acid (**129**), 1-pentadeca-7-10-dienyl-1,3-dihydroxy benzene (**130**), 1-pentadeca-8-enyl-3,3-dihydroxybenzene (**131**), I-4¹,II-3¹,4¹,I-5,II-5,I-7-hexahydroxy[I-3,II-8]biflavonone (**101**), I-4¹,II-4¹,I-5,II-5,I-7,II-7-hexahydroxy[I-3,II-8]biflavonone (3¹,8-binaringenin) (**102**), I-4¹,II-4¹,I-7,II-7-tetrahexahydroxy-[I-3,II-8]biflavonone (3¹,8-binaringenin) (**103**), *E*(-3-(pentadec-8-enyl)benzene-1,2-diol (**104**), 3-((7*E*,10*E*)-pentadeca-7,10-dien-1-yl)benzene-1,2-diol (**105**) from nuts of *Semecarpus anacardium* L.f.

Later, NS, Prakash Rao and Row LR [74] investigated more than seven compounds and two major compounds such as 1-pentadeca-7-10-dienyl-1,3-dihydroxy benzene (**130**) (Figure 20) and 1-penta deca-8-enyl-3,3-dihydroxybenzene (**131**) (Figure 20). Three biflavonoids were also isolated from the ethanol soluble fractions of *Semecarpus anacardium* nut shell and they characterized as I-4¹,II-3¹,4¹,I-5,II-5,I-7-hexahydroxy [I-3,II-8] biflavonone (**101**) (Figure 12). Gadam [75] reinvestigated that of bhilawanol (**128**) comprised of two components such as 1,2-dihydroxy-3-pentadecyl benzene (*E*(-3-(pentadec-8-enyl) benzene-1,2-diol (**104**) (Figure 21) and its corresponding diene analogue (3-((7*E*, 10*E*)-pentadeca-7,10-dien-1-yl)benzene-1,2-diol (**105**) (Table 2).

3.3.1.2. Compounds in Nuts and Leaves of Semecarpus anacardium L.f.

Ishratulla and co-workers [76] isolated tetrahydrorobustaflavone (**132**) (Figures 20, 22), tetrahydroamentoflavone (**106**) (Figures 12, 22) from the defatted nuts of the *Semecarpus anacardium* and the structure was characterized and also they reported amentoflavone (**36**) (Figures 12, 22), tetrahydroamentoflavone (**106**) as the sole compound from *Semecrpus anacardium* leaves.

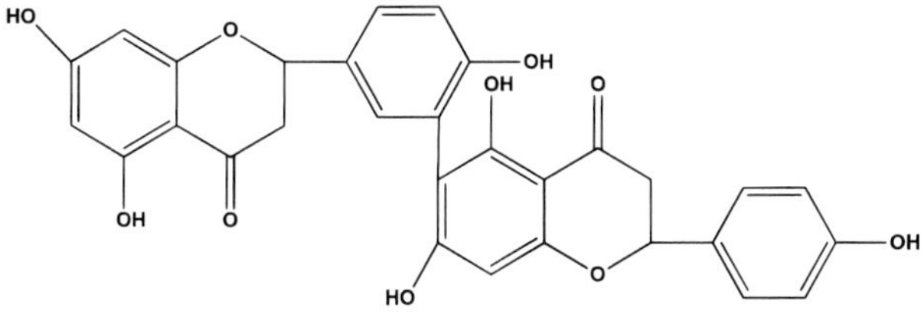

tetrahydrorobustaflavone (**132**)

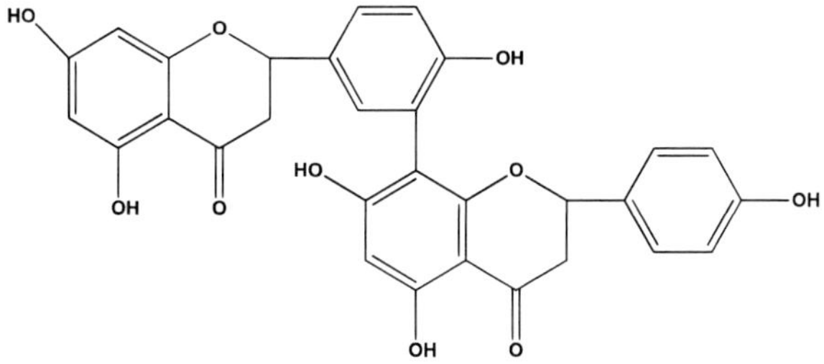

tetrahydroamentoflavone (**106**)

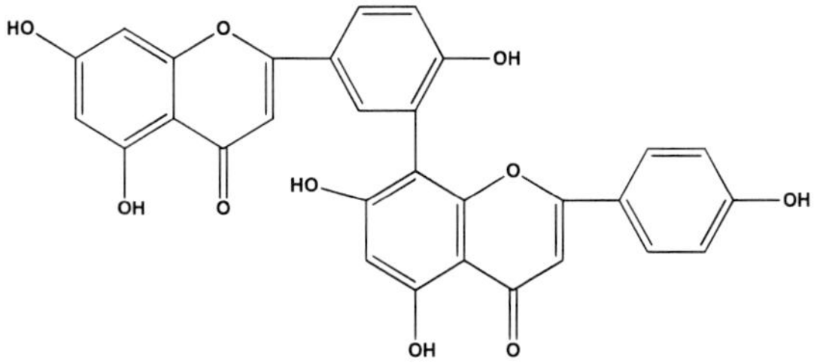

amentoflavone (**36**)

Figure 22. Tetrahydrorobustaflavone (**132**) (Figure 20), tetrahydroamentoflavone (**106**), amentoflavone (**36**) from nuts and leaves of *Semecarpus anacardium*.

3.3.1.3. Compounds in Nutshells of Semecarpus anacardium L.f.

Murthy SSN isolated three other biflavonoids, jeediflavanone (**108**) [77, 78], semicarpouflavonone (**109**) [79], galluflavanone (**110**) [80] from the alcoholic fraction of nutshells and were characterized (Figures 12, 23). It showed the presence of two chelated hydroxyl groups.

Latter, the isolated other two new dimeric biflavanoids, nallaflavonone (**111**), semicarpetine (**133**) (Figure 23) and anacarduflavanone (**134**) (Figure 23) from the nutshell and characterized.

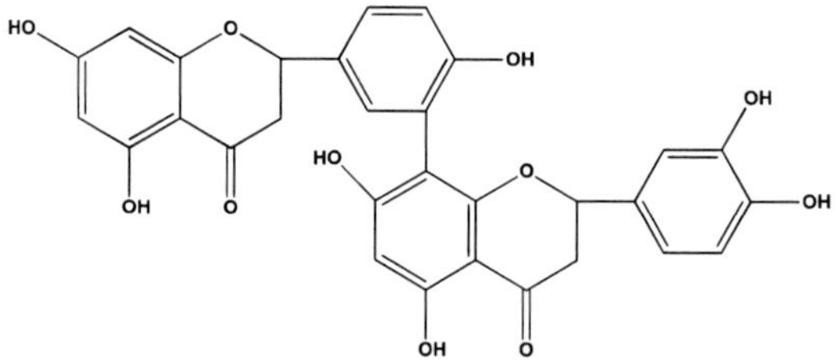

jeediflavanone (**108**)

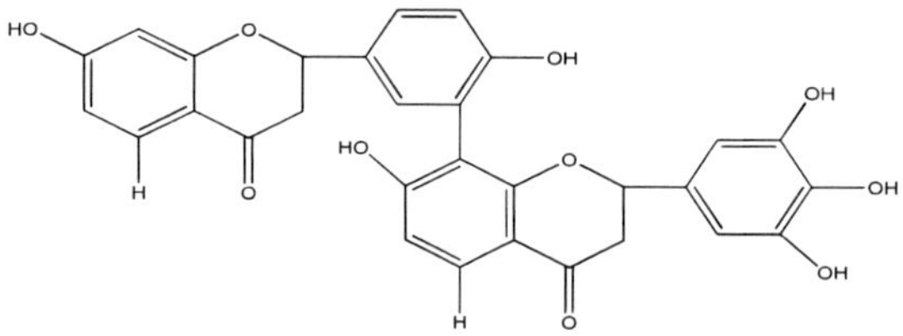

semicarpouflavonone (**109**)

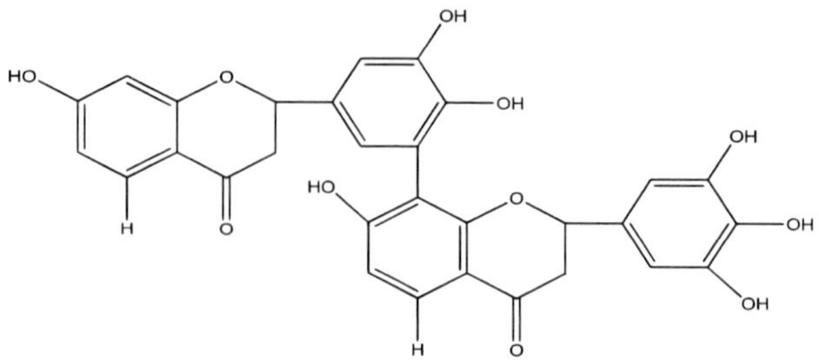

galluflavanone (**110**)

Figure 23. (Continued).

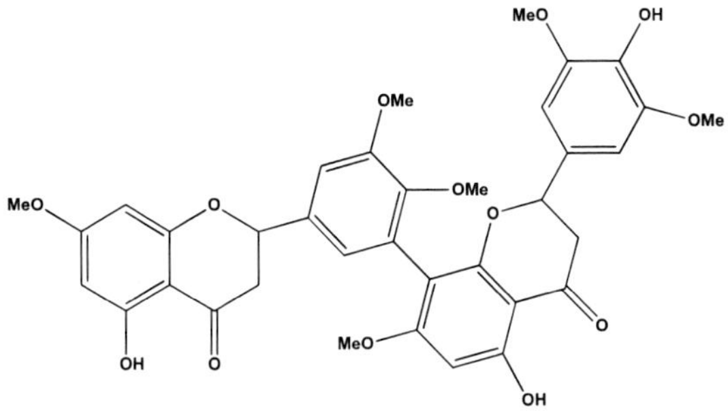

Nature-Inspired Phytochemicals ... 187

nallaflavonone (**111**)

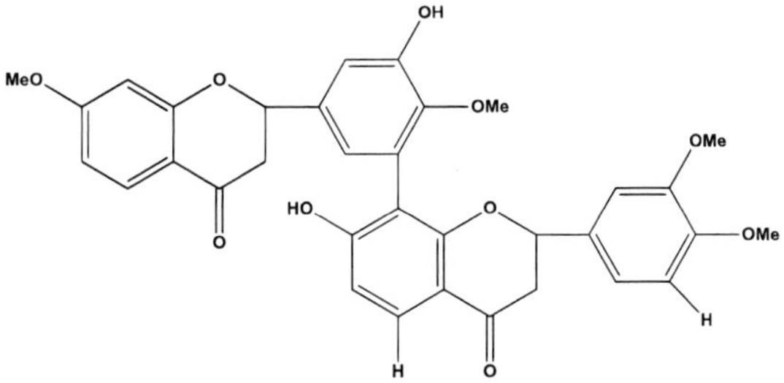

semecarpetine (**133**)

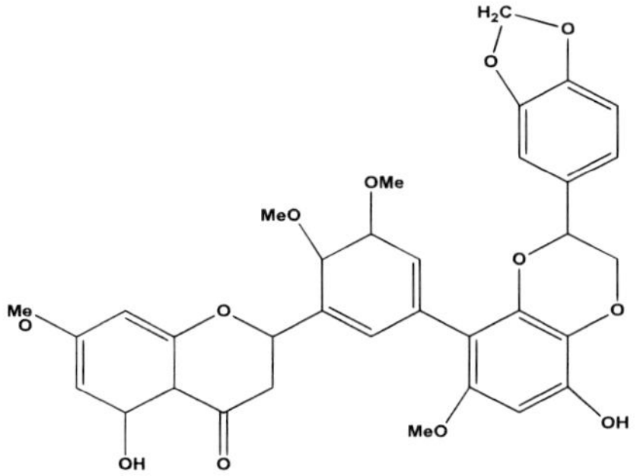

anacarduflavanone (**134**)

Figure 23. Jeediflavanone (**108**), semicarpouflavonone (**109**), galluflavanone (**110**) (Figure 12), nallaflavonone (**111**) (Figure 12), semecarpetine (**133**), anacarduflavanone (**134**) from nutshells of *Semecarpus anacardium*.

3.3.1.4. Compounds in Seeds of Semecarpus anacardium L.f.

In 1995, a new phenolic glucoside, anacardoside (**135**) (Figure 24) was isolated by Ramakumar and co-workers [84], and its structure as well as configuration were elucidated by a combination of NMR techniques as 1-*O*-β-D-glucopyranosyl-(1-6)-β-D-glucopyranosyloxy-3-hydroxy-5-methyl benzene.

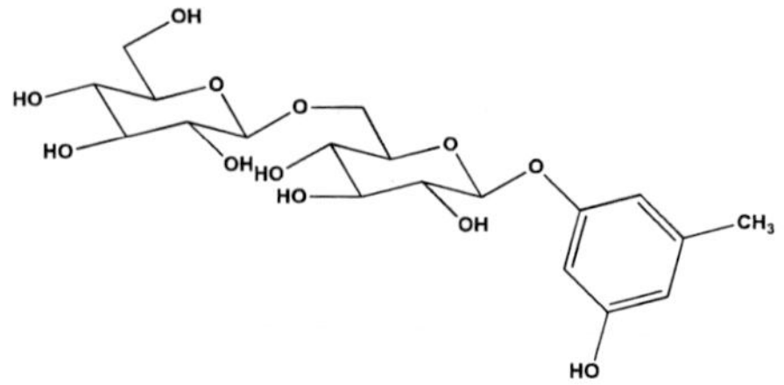

anacardoside (**135**)

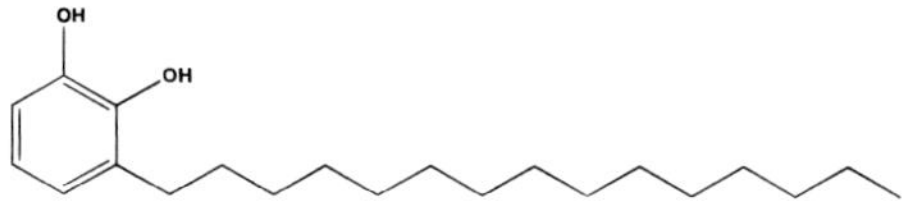

3-pentadecylbenzene (**136**)

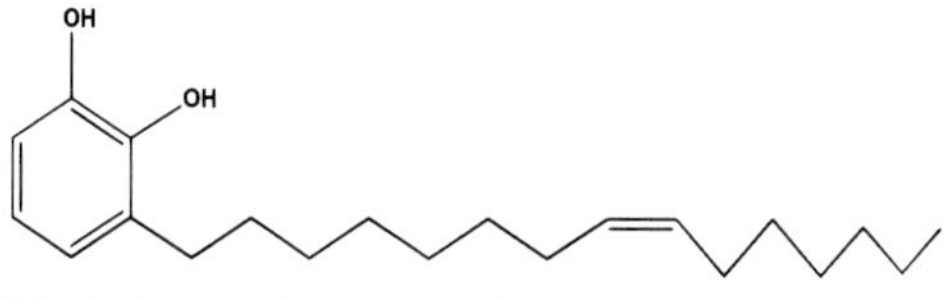

(Z)-3-(pentadec-8-enyl)benz-1,2-diol (**137**)

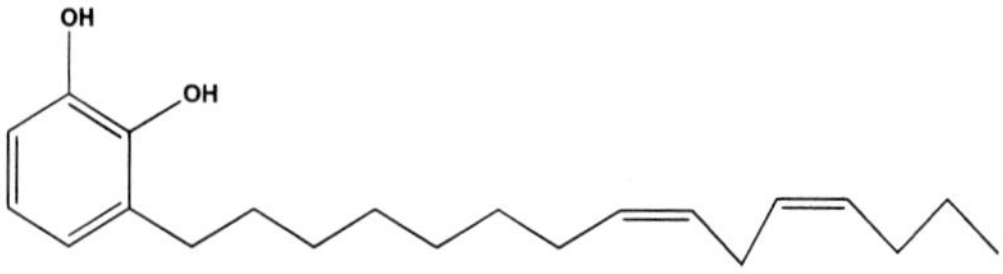

3-((8Z,11Z)-pentadeca-8,11-dienyl)benzene-1,2-diol (**138**)

Figure 24. Anacardoside (**135**), 3-pentadecylbenzene (**136**), (Z)-3-(pentadec-8-enyl)benz-1,2-diol (**137**), 3-((8Z,11Z)-pentadeca-8,11-dienyl)benzene-1,2-diol (**138**) from seeds of *Semecarpus anacardium*.

3-Pentadecylbenzene (**136**), and (Z)-3-(pentadec-8-enyl)benz-1,2-diol (**137**) (Figure 24), along with minor amount of saturated bhilawanol (**128**) (Figure 21) and 3-((8Z,11Z)-pentadeca-8,11-dienyl)benzene-1,2-diol (**138**) (Figure 24) were reported by Nagabhushana KS and co-workers [85].

3.3.1.5. Compounds in Stembark of Semecarpus anacardium L.f.

3,4,2[1],4[1]-Tetrahydroxychalcone (**139**) and 7,3[1],4[1]-trihydroxyflavone (**140**) (Figure 25) from ethyl acetate extract of *Semecarpus anacardium*L.f. stembark [86].

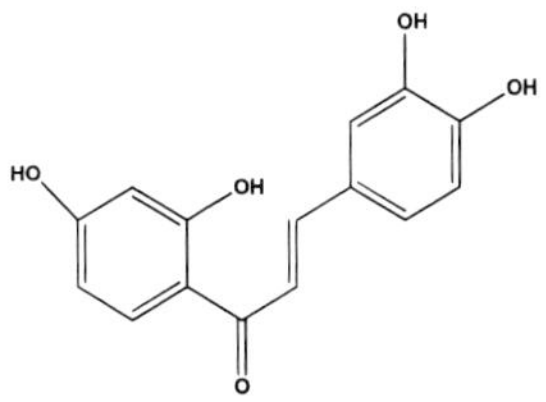

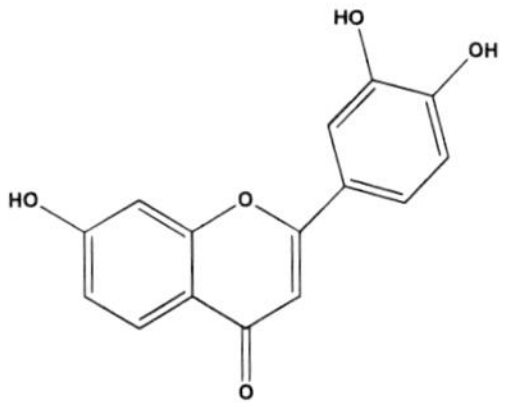

Nature-Inspired Phytochemicals ... 189

3,4,2¹,4¹-tetrahydroxychalcone (**139**) 7,3¹,4¹-trihydroxyflavone (**140**)

Figure 25. 3,4,2¹,4¹-Tetrahydroxychalcone (**139**) and 7,3¹,4¹-trihydroxyflavone (**140**) from stembark of *Semecarpus anacardium*.

3.3.2. Pharmacological Activities of Semecarpus anacardium L.f. (Figure 26)

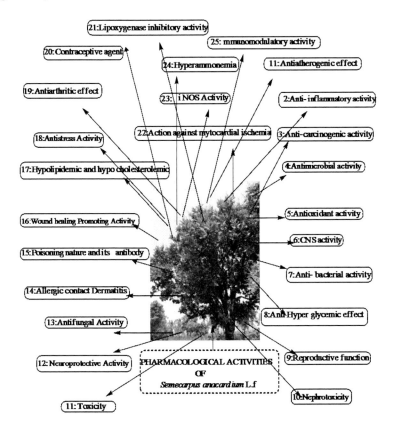

Figure 26. Phamacological activities of *Semecarpus anacardium* L.f.

Table 3. Phamacological activities of *Semecarpus anacardium (Sa)* L.f.

Plant part	Plant extract	Author (s)	Pharmacological activity	In Vitro/In Vivo	Ref.
nutshell	nut milk extract	Sharma *et al.*	1. Antiatherogenic effect (preventing atherogenesis)	It inhibits peroxidation of lipids in low density lipoproteins (*in vivo*)	[87]
nuts	nut extract	Ramprasath *et al.*	2. Anti-inflammatory activity: It means that the property of a substance or treatment that reduces inflammation or swelling	It decreases the carrageenan-induced paw edema and cotton pellet granuloma (*in vivo*)	[88]
	ethyl acetate extract of nuts	Salvem *et al.*		It shows dose-dependently anti-inflammatory effect of tetrahydroamentoflavone (THA) (*in vitro*)	[89]
fruits	methanolic, ethanolic, chloroform, ethyl acetate and petroleum ether extract of fruits	Bhitre *et al.* Satyavati *et al.*		Studied using the technique of carrageenan-induced paw oedema in albino rats (*in vivo*)	[90]
				Showed the anti-inflammatory activity of both immunological and non-immunological origin	[91]
nuts	crude ethanolic extract	Singh D *et al.*		Inhibited the spontaneous and lipopolysaccharide (LPS)-induced production of proinflammatory cytokines (*in vitro*)	[92]
stembark	ethyl acetate extract	Premlatha *et al.*		Informed the immunomodulatory potency and detoxification of a potent hepatocarcinogen, aflatoxin	[93]
nuts	nut milk extract	Ramprasath VR		It recouped the altered anti oxidant defense components to near normal levels	[94]
	nut milk extract and fresh dried powder of *Emblica officinalis* fruit with honey	Mythilipriya *et al.*		Showed its potent antioxidant analgesic, antipyretic, non ulcerogenic properties (*in vivo*)	[95]
	nut extract	Mathivadhani P *et al.*	3. Anti-carcinogenic activity:(inhibits the development of cancer)	Cytotoxicity analysis suggested that these cells had undergone changes due to the biochemical activity of SA nut extract and finally cells become apoptotic (*in vitro*)	[95]

Plant part	Plant extract	Author(s)	Pharmacological activity	In Vitro/In Vivo	Ref.
nut milk	Sa nut milk extract, dried powder of Phyllanthus embelica fruit and honey	Arulkumaran S et al.		Showed normal lipid peroxidation antioxidant defenses in mitochondrial enzymes (in-vivo)	[96]
	Sa nut milk extract	Sugapriya D et al.		Resulted the clearance of the leukemik cells from the Bone marrow and internal organs	[97]
	nut milk extract	Arathi G et al.		It significantly lowered the glycolytic activities of the enzymes and there was a rising gluconeogenic enzymes (in vivo)	[98]
		Periasamy VS et al. Sowmya-lakshmi S		Showed the changes in the nuclear morphology and cytoplasmic features (in vitro)	[99] [100]
nuts	n-hexane and chloroform extract	Sowmya-lakshmi S		Significantly active against MDA 231 cells than MCF-7 cells	[100]
	methanolic extract	Patel SR et al.		It shows its action on humam epidermoid larynx carcinoma cell line	[101] [102]
	petroleum ether and aqueous fractions	Mohanta et al.	4. Antimicrobial activity: (inhibits the growth of microorganisms)	Showed inhibition against Bacillus licheniformis, Vibrio cholarae, Pseudomonas aeruginosa	[102]
nuts,leaves, twings and green fruit	alcoholic extract	Nair A et al.	It showed bactericidal activity in vitro against 3 Gram-negative strains and 2 Gram-positive strains		[103]
nuts	aqueous extract	Verma N et al.	5. Antioxidant: (molecule that inhibits the oxidation of other molecules)	Increase in the activities of antioxidant enzymes, where as lactate dehydrogenase (LDH) activity is brought down significantly indicating a decrease in carcinogenesis	[104]

Table 3. (Continued)

Plant part	Plant extract	Author (s)	Pharmacological activity	In Vitro/In Vivo	Ref.
stembark	ethyl acetate extract	Sahoo AK et al.		It exhibits the antioxidant and anti apoptotic effects of Amruthaballi, Kalpamrutha to protect from cardiovascular disease (CVD) (*in vitro*)	[105]
nuts	milk extract	Farooq SM et al.	6. Central nervous system (CNS) activity: which coordinates theactivity of the entire nervous system - compare peripheral nervous system	It shows locomotor and nootropic activities	[106]
	juice of pericacrp	Amaley kranthikumar	7. Anti-bacterial activity: (destruction of bacteria)	Shown marked bacterial activity against Bacillus pyogens, *Escherichia coli*, Staphylococcus, Streptococcus	[107]
	ethanolic extract	Arul B et al.	8. Anti-hyper glycemic effect: is a condition in which an excessive amount of glucose circulates in the blood plasma	Reduces the blood glucose levels of normal rats (*in vivo*)	[108]
	SemecarpusL.f, *Embelicaofficinalis* and honey			Increased levels of total cholesterol, free cholesterol, phospholipids, triacylglycerides and free fatty acids and decreased levels of ester cholesterol in plasma, liver and kidney found in cancer suffering animals	[109]
	Sa extract	Sharma A et al.	9. Reproductive function: The organs and glands in the body that aid in the production of new individuals	Reduction in number of spermatogenic cells and spermatozoa in male albino rats. It also revealed the alteration in androgen levels in testis may effect the transformation of spermatocides to spermatids	[110]

Plant part	Plant extract	Author (s)	Pharmacological activity	In Vitro/In Vivo	Ref.
stembark	methanolic extract	Vinutha B et al.		It showed acetylcholinesterase inhibitory activity	[111]
nuts	*Sa* nut oil extract	Choudhari CV et al.	10. Nephrotoxicity: It is one of the most common kidney problems and occurs when your body is exposed to a drug or toxin that causes damage to your kidneys. When kidney damage occurs, you are unable to rid your body of excess urine, and wastes	It showed decrease in haemoglobin present and lowering of erythrocytes, indicating anemia during toxicity studies	[112] [113]
		Prabhu D et al.		Antimutagenic effect was seen (*in vivo*)	[114]
medicinal plants of Anacardiceae family.	aqueous extract	Krishnarajua AV et al.		Significant cytotoxicity was observed	[115]
sap, nut	Oil	Matthai TP	11. Toxicity: degree to which a substance can damage an organism	The oily part of nut is toxic and its degree of removal is proportional to its safety margin	[116]
fruits	fruit resin extract	Adhami HR et al.	12. Neuroprotective activity: refers to the relative preservation of neuronal structure	It show sacetylcholinesterase inhibitory activity	[117]
nuts	methanolic extract	Matthai TP et al.	13. Antifungal-activity: It prevents fungal growth	The excellent inhibitory activity was observed against *Rhizoctonia solanii* followed by *Sclerotium rolfsii* etc. due to the presence of alkaloids, saponins, tannins, flavanoids, steroids, glycosides	[118]
seeds	burning of seeds	Khailashbharathi et al.	14. Allergic contact dermatitis: It is also called contact allergy	The smoke produced by its burning cause allergic contact dermatitis over a time	[119]

Table 3. (Continued)

Plant part	Plant extract	Author (s)	Pharmacological activity	In Vitro/In Vivo	Ref
nuts	milk extract	Vijayalakshmi et al.	15. Poisoning nature and its antidote: an agent that counteracts a poison.	It showed moderate increase in the levels of blood glucose, plasma urea, uric acid, and creatinine and lipid alteration	[120]
stembark	methanolic extract		16. Wound healing promoting activity: It is the process by which skin or other body tissue repairs itself after trauma	Epithelialization of the incision wound was faster with a high rate of wound contraction. The tensile strength of the incision wound was significantly increased. There was an increased cross linking of collagen fibers and absence of monocytes as compared to control	[121]
nuts		Tripathi YB et al.	17. Hypolipidemic and hypocholesterolemia due to a genetic defect in the receptor	Controls the tissue metabolism and improving the level of insulin secretion and action. The increased insulin output in diabetic animals lowers the intestinal absorption of cholesterol and enhancing the extraction of ingested cholesterol and inhibition of 3-hydroxy-3-methylglutaryl-CoA (HMG Co-A) reductase could contribute to the hypolipidemic activity	[112]
		Vijayalakshmi et al.		They show their reactivity towards reactive oxygen species ability to chelate transition metallic cations and regenerate α-tocopherol	[120]
	nut extract	Shukkla SD et al.	18. Antistress: activity acting to prevent or reduce stress	Reduced the generating cell bodies and it shows neuroprotective effect	[123]
	milk extract	Satyavathi GV et al. Vijayalakshmi T et al.	19. Antiarthritic effect: an agent used in treatment of arthritis	Significant increase in the levels of lipid peroxides, reactive oxygen species (ROS), reactive nitrogen species (RNS) to be significantly decreased on administration of the drug	[91, 95], 124]

Plant part	Plant extract	Author (s)	Pharmacological activity	In Vitro/In Vivo	Ref
nuts	nut milk extract	Nada R et al.		It brings down the increased levels of lysosomal enzymes in adjuvant induced arthritis rats and good therapeutic agent for the arthritis	[125]
aerial parts of Sa	water extract	Narayan et al.	20. Contraceptive agent: a form of birth control which prevents the sperm from fertilising the egg is a contraceptive agent	Exhibits spermicidal activity	[126]
fruit	ethanolic extract			Significant reduction in the sperm motility and density was observed	
nuts		Nagabhushana et al.	21. Lipoxygenase inhibitory activity: are a family of (non-heme), iron-containing enzymes most of which catalyze the dioxygenation of poly-unsaturated fatty acids in lipids containing a cis-1,4-pentadiene into cell signaling agents that serve diverse roles as autocrine signals that regulate the function of their parent cells, paracrine signals that regulate the function of nearby cells, and endocrine signals that regulate the function of distant cells	Potent inhibitor of both soyabean and potato lipoxygenases	[85]
fruits	ethanolic extract	Prasannakur SR et al.	22. Action against myocardial ischemia: Myocardial ischemia is an intermediate condition in coronary artery disease during which the heart tissue is slowly or suddenly starved of oxygen and other nutrients nutrients	Provide significant recovery in developing tension and heart rate from ischemia	[127]

Table 3. (Continued)

Plant part	Plant extract	Author (s)	Pharmacological activity	In Vitro/In Vivo	Ref
fruits	ethanolic extract	Prasannakur SR et al.	22. Action against myocardial ischemia: Myocardial ischemia is an intermediate condition in coronary artery disease during which the heart tissue is slowly or suddenly starved of oxygen and other nutrients nutrients	Provide significant recovery in developing tension and heart rate from ischemia	[127]
nuts	oily fraction	Tripathi B et al.	23. Inducible nitric oxide synthase (iNOS) activity	Enhances the expression of protein kinase C (PKC protein), which may be responsible for its reported pro-inflammatory property	[128]
	milk extract	Vijaykumar et al.	24. Hyperammonemia: is a metabolic disturbance characterized by an excess of ammonia in the blood	Against ammonium chloride-induced hyperammonemia	[129]
	milk extract	Balachandran et al.	25. Immunomodulatory activity: a chemical agent that modifies the immune response or the functioning of the immune system	Reduced immunoglobulin G (IgG) and elevated immunoglobulin A (IgA) and immunoglobulin M (IgM) in the hepatocellular carcinoma and also against Hepatitis B virus-induced hepatocellular carcinoma	[130]

CONCLUSION

Semecarpus anacardium L.f. is a well-known medicinal plant in both Ayurveda and Siddha medicines. It has been found to have lot of medicinal properties, particularly for its anticancerous activity. The present review deals with the distribution, phytochemical and pharmacological aspects of *Semecarpus anacardium* L.f. Plant improvement studies (seed germination and *in vitro* propagation) of *Semecarpus anacardium* L.f. are also discussed. More than 20 Biflavonoids and their biological activity studies have been reported from *Semecarpus anacardium* L.f till today. We have selected "*Semecarpus anacardium* L.f," for its high medicinal value in Ayurveda and Siddha systems and isolated several active constituents which is very well known as "Bhallataka". Isolation of phytochemical constituents from "Bhallataka"plant parts and synthesis of Bhallataka biodiesel, biolubricants, novel applications and their biological activities were done in our laboratory [131-142].

REFERENCES

[1] Nimberkar TP, Katolkar PP. Traditional knowledge of medicinal plants of Gondia district (In Maharashtra State): an ethno-botanical survey. *J Herbal Med Toxicol* 5(2), 9-17, 2011.

[2] Cragg GM, Newman DJ. Natural products: a continuing source of novel drug leads. *Biochim Biophys Acta* 1830(6), 3670-3695, 2013.

[3] Phillipson JD. Natural products as drugs. *Trans R Soc Trop Med Hyg* 88(Suppl 1), S17-S19, 1994.

[4] De Smet PA. The role of plant-derived drugs and herbal medicines in healthcare. *Drugs* 54(6), 801-840, 1997.

[5] Newman DJ, Cragg GM, Snader KM. Natural products as sources of new drugs over the period 1981-2002. *J Nat Prod* 66(7), 1022-1037, 2003.

[6] Benowitz S. As war on cancer hits 25 years mark, scientists see progress, challenges. *The Scientist* 10, 127, 1996.
[7] Hamburger M, Hostettmann K. Bioactivity in plants: the link between phytochemistry and medicine. *Phytochemistry* 30(12), 3864-3874, 1991.
[8] Phillipson JD. Phytochemistry and medicinal plants. *Phytochemistry* 56(3), 237-243, 2001.
[9] Kong JM, Goh NK, Chia LS, Chia TF. Recent advances in traditional plant drugs and orchids. *Acta Pharmacol Sin* 24(1), 7-21, 2003.
[10] Kumar GR, Kumar R. Bridging traditional medicines with modern biotechnology. *Chemtracts* 15(13), 693-705, 2002.
[11] Phillipson JD. 50 years of medicinal plant research - every progress in methodology is a progress in science. *Planta Med* 69(6), 491-495, 2003.
[12] Chun KS, Kang JY, Kim OH, Kang H, Surh YJ. Effects of yakuchinone A and yakuchinone B on the phorbol ester-induced expression of COX-2 and iNOS and activation of NF-kappaB in mouse skin. *J Environ Pathol Toxicol Oncol* 21(2), 131-139, 2002.
[13] Wen J, You KR, Lee SY, Song CH, Kim DG. Oxidative stress-mediated apoptosis. The anticancer effect of the sesquiterpene lactone parthenolide. *J Biol Chem* 277(41), 38954-38964, 2002.
[14] Youdin KA, Deans SG, Finlayson HJ. The antioxidant properties of thyme (*Thymus zygis* L.) essential oil: an inhibitor of lipid peroxidation and a free radical scavenger. *J Essential Oil Res* 14(3), 210-215, 2002.
[15] Gören N, Arda N, Çaliskan Z. Chemical characterization and biological activities of the genus Tanacetum (Compositae). *Studies in Natural Products Chemistry* 7(Part H), 547-658, 2002.
[16] Heinrich M. Ethnobotany and natural products: the search for new molecules, new treatments of old diseases or a better understanding of indigenous cultures? *Curr Top Med Chem* 3(2), 141-154, 2003.
[17] Petersen M, Simmonds MS. Rosmarinic acid. *Phytochemistry* 62(2), 121-125, 2003.

[18] Hanrahan JR, Chebib M, Davucheron NL, Hall BJ, Johnston GA. Semisynthetic preparation of amentoflavone: A negative modulator at GABA(A) receptors. *Bioorg Med Chem Lett* 13(14), 2281-2284, 2003.

[19] Yoshikawa M, Xu F, Morikawa T, Ninomiya K, Matsuda H. Anastatins A and B, new skeletal flavonoids with hepatoprotective activities from the desert plant *Anastatica hierochuntica*. *Bioorg Med Chem Lett* 13(6), 1045-1049, 2003.

[20] Delporte C, Backhouse N, Salinas P, San-Martín A, Bórquez J, Loyola A. Pharmaco-toxicological study of diterpenoids. *Bioorg Med Chem* 11(7), 1187-1190, 2003.

[21] Kumar S, Gautam S, Sharma A. Antimutagenic and antioxidant properties of plumbagin and other naphthoquinones. *Mutat Res* 755(1), 30-41, 2013.

[22] Southwell IA, Russel MF. *Acta Horticulturare, 597 (Proceedings of the International Conference on Medicinal and Aromatic Plants)* Part-2, 2001, 31-47, 2003.

[23] Imbert TF. Discovery of podophyllotoxins. *Biochimie* 80(3), 207-222, 1998.

[24] Srivastava V, Negi AS, Kumar JK, Gupta MM, Khanuja SP. Plant-based anticancer molecules: a chemical and biological profile of some important leads. *Bioorg Med Chem* 13(21), 5892-5908, 2005.

[25] Noble RL. The discovery of the vinca alkaloids--chemotherapeutic agents against cancer. *Biochem Cell Biol* 68(12), 1344-1351, 1990.

[26] Jacoby M. Taxol. *Chem Eng News* 83(25), 120, 2005.

[27] Wani MC, Taylor HL, Wall ME, Coggon P, McPhail AT. Plant antitumor agents. VI. The isolation and structure of taxol, a novel antileukemic and antitumor agent from *Taxus brevifolia*. *J Am Chem Soc* 93(9), 2325-2327, 1971.

[28] Schiff PB, Fant J, Horwitz SB. Promotion of microtubule assembly *in vitro* by taxol. *Nature* 277(5698), 665-667, 1979.

[29] Wilson L, Jordan MA. Microtubule dynamics: taking aim at a moving target. *Chem Biol* 2(9), 569-573, 1995.

[30]. Kinghorn AD, Seo EK. Plants as sources of drugs. *Agricultural Materials as Renewable Resources* 647, 179-193,1996.
[31] Oberlies NH, Kroll DJ. Camptothecin and taxol: historic achievements in natural products research. *J Nat Prod* 67(2), 129-135, 2004.
[32] Cortes J, Baselga J. Targeting the microtubules in breast cancer beyond taxanes: the epothilones. *Oncologist* 12(3), 271-280, 2007.
[33] Hsiang YH, Hertzberg R, Hecht S, Liu LF. Camptothecin induces protein-linked DNA breaks *via* mammalian DNA topoisomerase I. *J Biol Chem* 260(27), 14873-14878, 1985.
[34] Christenhusz MJM, Byng JW. The number of known plants species in the world and its annual increase. *Phytotaxa* 261(3), 201–217, 2016.
[35] Correia SdeJ, David JP, David JM. Secundary metabolites from species of Anacardiaceae. *Quim Nova* 29(6), 1287-1300, 2006. In Spanish.
[36] Cardoso MP, David JM, David JP. A new alkyl phenol from *Schinopsis brasiliensis*. *Nat Prod Res* 19(5), 431-433, 2005.
[37] Weniger B, Vonthron-Sénécheau C, Arango GJ, Kaiser M, Brun R, Anton R. A bioactive biflavonoid from *Campnosperma panamense*. *Fitoterapia* 75(7-8), 764-767, 2004.
[38] Kumanotani J. Urushi (oriental lacquer). A natural aesthetic durable and future-promising coating, *Prog Org Coat* 26,163,1995.
[39] Wu PL, Lin SB, Huang CP, Chiou RY. Antioxidative and cytotoxic compounds extracted from the sap of *Rhus succedanea*. *J Nat Prod* 65(11), 1719-1721, 2002.
[40] Kubo I, Muroi H, Himejima M, Yamagiwa Y, Mera H, Tokushima K, Ohta S, Kamikawa T. Structure-antibacterial activity relationships of anacardic acids. *J Agric Food Chem* 41(6), 1016-1019, 1993.
[41] Gambaro V, Chamy MC, von Brand E, Garbarino JA. 3-(pentadec-10-enyl)-catechol, a new allergenic compound from *Lithraea caustica* (Anacardiaceae). *Planta Med* (1), 20-22, 1986.

[42] Rivero-Cruz JF, Chavez D, Bautista BH, Anaya AL, Matat R. Separation and characterization of *Metopium brownei* urushiol components. *Phytochemistry* 45(5), 1003-1008, 1997.
[43] David JM, Chávez JP, Chai HB, Pezzuto JM, Cordell GA. Two new cytotoxic compounds from *Tapirira guianensis*. *J Nat Prod* 61(2), 287-289, 1998.
[44] Queiroz EF, Kuhl C, Terreaux C, Mavi S, Hostettmann K. New dihydroalkylhexenones from *Lannea edulis*. *J Nat Prod* 66(4), 578-580, 2003.
[45] López P, Ruffa MJ, Cavallaro L, Campos R, Martino V, Ferraro G. 1,3-dihydroxy-5-(tridec-4',7'-dienyl)benzene: a new cytotoxic compound from *Lithraea molleoides*. *Phytomedicine* 12(1-2), 108-111, 2005.
[46] Valcic S, Wächter GA, Eppler CM, Timmermann BN. Nematicidal alkylene resorcinols from *Lithraea molleoides*. *J Nat Prod* 65(9), 1270-1273, 2002.
[47] Du Y, Oshima R, Yamauchi Y, Kumanotani J, Miyakoshiji T. Long chain phenols from the Burmese lac tree, *Melanorrhoea usitate*. *Phytochemistry* 25(9), 2211-2218, 1986.
[48] Adawadkar PD, Elsohly MA. An urushiol derivative from poison sumac. *Phytochemistry* 22(5), 1280-1281, 1983.
[49] Carpenter RC, Sotheeswaran S, Sultanbanwa MUS, Balasubramaniam S. (-)-5-Methylmellein and catechol derivatives from four *Semecarpus* species. *Phytochemistry* 19(3), 445-447, 1980.
[50] Rao NSP, Row LR, Brown RT. Phenolic constituents of *Semecarpus anacardium*. *Phytochemistry* 12(3), 671-681, 1973.
[51] Coates NJ, Gilpin,ML, Gwynn MN, Lewis DE, Milner PH, Spear SR, Tyler JW. A novel beta-lactamase inhibitor isolated from *Spondias mombin*. *J Nat Prod* 57(5), 654-657, 1994.
[52] Correia SD, David JM, David JP, Chai HB, Pezzuto JM, Cordell GA. Alkyl phenols and derivatives from *Tapirira obtusa*. *Phytochemistry* 56(7), 781-784, 2001.

[53] Mamdapur VR. Characterization of alkenylresorcinol in mango (*Mangifera indica* L.) latex. *J Agric. Food Chem* 33(3), 377-379, 1985.
[54] Sharma SK, Ali M. Chemical constituents of stem bark of *Mangifera indica* Linn. (cultivar desi). *J Indian Chem Soc* 72(5), 339-342, 1995.
[55] Ahmad I, Ishratullah K, Ilyas M, Rahman W, Seligmann O, Wagner H. Tetrahydroamentoflavone from nuts of *Semecarpus prainii*. *Phytochemistry* 20(5), 1169-1170, 1981.
[56] Murthy SSN. Jeediflavanone - a biflavonoid from *Semecarpus anacardium*. *Phytochemistry* 24(5), 1065-1069, 1985.
[57] Gil RR, Lin LZ, Cordell GA, Kumar MR, Ramesh M, Reddy BM, Mohan GK, Narasimha AV, Rao A. Anacardoside from the seeds of *Semecarpus anacardium*. *Phytochemistry* 39(2), 405-407, 1995.
[58] Lin YM, Chen FC. Robustaflavone from the seed-kernels of *Rhus succedanea*. *Phytochemistry* 13(8), 1617-1619, 1974.
[59] Lin YM, Anderson H, Flavin MT, Pai YH, Mata-Greenwood E, Pengsuparp T, Pezzuto JM, Schinazi RF, Hughes SH, Chen FC. *In vitro* anti-HIV activity of biflavonoids isolated from *Rhus succedanea* and *Garcinia multiflora*. *J Nat Prod* 60(9), 884-888, 1997.
[60] Masesane IB, Yeboah SO, Liebscher J, Mügge C, Abegaz BM. A bichalcone from the twigs of *Rhus pyroides*. *Phytochemistry* 53(8), 1005-1008, 2000.
[61] Ahmed MS, Galal AM, Ross SA, Ferreira D, ElSohly MA, Ibrahim AS, Mossa JS, El-Feraly FS. A weakly antimalarial biflavanone from *Rhus retinorrhoea*. *Phytochemistry* 58(4), 599-602, 2001.
[62] (Proceedings) Schulze-kaysers N, Feuereisen MM, Schieber A. Phenolic compounds in edible species of the Anacardiaceae family - a review. *RSC Advances* 5, 73301-73314, 2015.
[64] Chetty KM, Sivaji K, Rao KT. *Flowering Plants of Chittoor District, Andhra Pradesh, India*. - Students offset printers, Tirupati, pp.76, 131, 2008.

[65] Maheshwari JK. *Addition to the Flora of Delhi.* p.48. Department of Botany, University of Delhi, 1996.
[66] Singh NP. *Flora of Eastern Karnataka.* Vol. 1. pp. 46, 61, 81, 98, 100, 117, 121, 207, 210, Mittal Publications, 1988.
[67] kumar A. A review on essential perspectives for *Semecarpus anacardium. International Journal of Pharmaceutical and Chemical Sciences* 3(1), 225-230, 2014.
[68] Poornima M, Anitha S, kumar GV, *Semecarpus anacardium*: a review, *International Ayurvedic Medical Journal (IJMJ)* 1(6), 69-76, 2013.
[69] Panda BM, Hazra S. Seedling culture of *Semecarpus anacardium* L. *Seed Science and Biotechnology* 3(2), 54-59, 2009.
[70] Naidu DS. Constituents of the marking-nut: *Semecarpus anacardium* Linn. *J Indian Inst Sci* 8, 129-142, 1925.
[71] Pillay P, Siddiqui S. Chemical investigation of marking nut. *J Indian Chem Soc* 8, 1517, 1931.
[72] Chattopadyaya MK, Khare RL. Isolation of anacardic acid from *Semecarpus anacardium* L.f. and study of its anthelmintic activity. *Indian Journal of Pharmacy* 31(4), 104-105, 1969.
[73] Govindachary TR, Joshi BS, Kamal VM, Phenolic constituents of *Semecarpus anacardium. Indian J Chem* 9, 1044-1046, 1971.
[74] Rao NSP, Row LR, Brown RT. Phenolic constituents of *Semecarpus anacardium. Phytochemistry* 12(3), 671-681, 1973.
[75] Gedam PH, Sampathkumaran PS, Sivasamban MA, Composition of bhilawanol from *Semecarpus anacardium. Phytochemistry* 13(2), 513–515, 1974.
[76] Ishatulla K, Ansari WH, Rahman W, Okigawa M, Kawanon N. Bioflavanoids from *Semecarpus anacardium* linn. *Indian J Chem* 15, 617–622, 1977.
[77] Murthy SSN. Jeediflavanone-a biflavonoid from *Semecarpus anacardium. Phytochemistry* 25(5), 1065-1069, 1985.
[78] Murthy SSN, Confirmation of the structure of jeediflavanone: A biflavanone from *Semecarpus anacardium. Phytochemistry* 23(4), 925-927, 1984.

[79] Murthy SSN. A biflavonoid from *Semecarpus anacardium*. *Phytochemistry* 22(6), 1518-1520, 1983.
[80] Murthy SSN. A biflavanone from *Semecarpus anacardium*. *Phytochemistry* 22(11), 2636-2638, 1983.
[81] Murthy SSN. New dimeric flavone from *Semecarpus anacardium*. *Proc Indian Natl Sci Acad* 57, 632, 1987.
[82] Murthy SSN. Semecarpetin, a biflavanone from *Semecarpus anacardium*. *Phytochemistry* 27(9), 3020–3022,1988.
[83] Murthy SS. New biflavonoid from *Semecarpus anacardium* linn. *Clin Acta Turcica* 20, 30-37, 1992.
[84] Gil RR, Lin LZ, Cordell GA, Kumar MR, Ramesh M, Reddy BM, Mohan GK, Narasimha AV, Rao A. Anacardoside from the seeds of *Semecarpus anacardium*. *Phytochemistry* 39(2), 405-407, 1995.
[85] Nagabhushana KS, Umamaheshwari S, Tocoli FE, Prabhu SK, Green IR, Ramadoss CS. Inhibition of soybean and potato lipoxygenases by bhilawanols from bhilawan (*Semecarpus anacardium*) nut shell liquid and some synthetic salicylic acid analogues. *J Enzyme Inhib Med Chem* 17(4), 255-259, 2002.
[86] Linn. Selvam C, Jachak SM, Bhutani KK. Cyclooxygenase inhibitory flavonoids from the stem bark of *Semecarpus anacardium*. *Phytother Res* 18(7), 582-584, 2004.
[87] Sharma A, Mathur R, Dixit VP. Hypocholesterolemic activity of nut shell extract of *Semecarpus anacardium* (Bhilawa) in cholesterol fed rabbits. *Indian J Exp Biol* 33(6), 444-448, 1995.
[88] Ramprasath VR, Shanthi P, Sachdanandam P. Immunomodulatory and anti-inflammatory effects of *Semecarpus anacardium* LINN. Nut milk extract in experimental inflammatory conditions. *Biol Pharm Bull* 29(4), 693-700, 2006.
[89] Selvam C, Jachak SM. A cyclooxygenase (COX) inhibitory biflavonoid from the seeds of *Semecarpus anacardium*. *J Ethnopharmacol* 95(2-3), 209-212, 2004.
[90] Bhitre MJ, Patil S, Kataria M, Anwikar S, Kadri H. Antiinflammatory activity of the fruits of *Semecarpus anacardium* Linn. *Asian J Chem* 20(3), 2047-2050, 2008.

[91] Satyavati GV, Prasad DN, Das PK, Singh HD. Anti-inflammatory activity of *Semecarpus anacardium* Linn, A preliminary study. *Indian J Physiol Pharmacol* 13(1), 37-45, 1969.
[92] Singh D, Aggarwal A, Mathias A, Naik S. Immunomodulatory activity of *Semecarpus anacardium* extract in mononuclear cells of normal individuals and rheumatoid arthritis patients. *J Ethnopharmacol* 108(3), 398-406, 2006.
[93] Premalatha B, Sachdanandam P. Potency of *Semecarpus anacardium* Linn. nut milk extract against aflatoxin B(1)-induced hepatocarcinogenesis: Reflection on microsomal biotransformation enzymes. *Pharmacol Res* 42(2), 161-166, 2000.
[94] Ramprasath VR, Shanthi P, Sachdanandam P. *Semecarpus anacardium* Linn. nut milk extract, an indigenous drug preparation, modulates reactive oxygen/nitrogen species levels and antioxidative system in adjuvant arthritic rats. *Mol Cell Biochem* 276(1-2), 97-104, 2005.
[95] Mathivadhani P, Shanthi P, Sachdanandam P. Apoptotic effect of *Semecarpus anacardium* nut extract on T47D breast cancer cell line. *Cell Biol Int* 31(10), 1198-1206, 2007.
[96] Arulkumaran S, Ramprasath VR, Shanthi P, Sachdanandam P. Alteration of DMBA-induced oxidative stress by additive action of a modified indigenous preparation--Kalpaamruthaa. *Chem Biol Interact* 167(2), 99-106, 2007.
[97] Sugapriya D, Shanthi P, Sachdanandam P. Restoration of energy metabolism in leukemic mice treated by a siddha drug--*Semecarpus anacardium* Linn. nut milk extract. *Chem Biol Interact* 173(1), 43-58, 2008.
[98] Arathi G, Sachdanandam P. Therapeutic effect of *Semecarpus anacardium* Linn. nut milk extract on carbohydrate metabolizing and mitochondrial TCA cycle and respiratory chain enzymes in mammary carcinoma rats. *J Pharm Pharmacol* 55(9), 1283-1290, 2003.
[99] Periasamy VS, Subash-Babu P, Muthukumaran VR, Akbarsha MA, Alshatwi AA. *In vitro* cytotoxic effect of formulated Semecarpus

ghee nanoemulsion on human cervical cancer (SiHa) cells. *Advanced Science Letters* 6(1), 75-79, 2012.

[100] Sowmyalakshmi S, Nur-E-Alam M, Akbarsha MA, Thirugnanam S, Rohr J, Chendil D. Investigation on *Semecarpus Lehyam*--a Siddha medicine for breast cancer. *Planta* 220(6), 910-918, 2005.

[101] Patel SR, Suthar AP, Patel RM. *In vitro* cytotoxicity activity of *Semecarpus anacardium* extract against Hep 2 cell line and vero cell line. *International Journal of PharmTech Research* 1(4), 1429-1433, 2009.

[102] Mohanta TK, Patra JK, Rath SK, D. K. Pal DK, Thatoi HN. Evaluation of antimicrobial activity and phytochemical screening of oils and nuts of *Semicarpus anacardium* L.f. *Scientific Research and Essay* 2(11), 486-490, 2007.

[103] Nair A, Bhide SV. Antimicrobial properties of different parts of *Semecarpus anacardium*. *Indian Drugs* 33(7), 323-328,1996.

[104] Verma N, Vinayak M. Immediate publication. *Bioscience Reports* p.BSR20080035, 2008.

[105] Sahoo AK, Narayanan N, Sahana S, Rajan SS, Mukherjee PK. *In vitro* antioxidant potential of *Semecarpus Anacardium* L. *Pharmacologyonline* 3, 327-335, 2008.

[106] Farooq SM, Alla TR, Rao NV, Prasad K, Shalam, Nandakumar K, Gouda TS, Satyanarayana S. A study on CNS effects of milk extract of nuts of *Semecarpus Anacardium*. Linn, (Anacardiaceae). *Pharmacologyonline* 1, 49-63, 2007.

[107] Amaley KD, Jain AS. A review article on Bhallataka (*Semecarpus Anacardium*. Linn). *International Ayurvedic Medical Journal* (*IAMJ*) 3(1), 1-6, 2015.

[108] Arul B, Kothai R, Christina AJ. Hypoglycemic and antihyperglycemic effect of *Semecarpus anacardium* Linn in normal and streptozotocin-induced diabetic rats. *Methods Find Exp Clin Pharmacol* 26(10), 759-762, 2004.

[109] Veena K, Shanthi P, Sachdanandam P. The biochemical alterations following administration of Kalpaamruthaa and *Semecarpus*

anacardium in mammary carcinoma. *Chem Biol Interact* 161(1), 69-78, 2006.

[110] Sharma A, Verma PK, Dixit VP. Effect of *Semecarpus anacardium* fruits on reproductive function of male albino rats. *Asian J Androl* 5(2), 121-124, 2003.

[111] Vinutha B, Prashanth D, Salma K, Sreeja SL, Pratiti D, Padmaja R, Radhika S, Amit A, Venkateshwarlu K, Deepak M. Screening of selected Indian medicinal plants for acetylcholinesterase inhibitory activity. *J Ethnopharmacol* 109(2), 359-363, 2007.

[112] Choudhari CV, Deshmukh PB. Acute and subchronic toxicity study of *Semecarpus anacardium* on hemoglobin percent and RBC count of male Albino rat. *J Herb Med Toxicol* 1, 43-45, 2007.

[113] Choudhari CV, Deshmukh PB. Effect of *Semecarpus anacardium* pericarp oil extract on histology and some enzymes of kidney in Albino rat. *J Herbal Med Toxicol* 2(1), 27-32, 2008.

[114] Prabhu D, Rajwani LS, Desai PV. The antimutagenic effect of *Semecarpus anacardium* under *in vivo* condition. *Asian J Chem* 12, 13-16, 2005.

[115] Krishnaraju AV, Rao TVN, Sundararaju D, Vanisree M, Tsay HS, Subbaraju GV. Assessment of bioactivity of Indian medicinal plants using brine shrimp (*Artemia salina*) lethality assay. *International Journal of Applied Science and Engineering* 3(2), 125-134, 2005.

[116] Matthai TP, Date A. Renal cortical necrosis following exposure to sap of the marking-nut tree (*Semecarpus anacardium*). *Am J Trop Med Hyg* 28(4), 773-774, 1979.

[117] Adhami HR, Linder T, Kaehlig H, Schuster D, Zehl M, Krenn L. Catechol alkenyls from *Semecarpus anacardium*: acetylcholinesterase inhibition and binding mode predictions. *J Ethnopharmacol* 139(1), 142-148, 2012.

[118] Jain P, Singh SK, Sharma HP, Basri F. Phytochemical screening and antifungal activity of *Semcarpus anacardium* L. (An anti-cancer plant). *International Journal of Pharmaceutical Sciences and Research* 5(5), 1884-1891, 2014.

[119] Bhatia K, Kataria R, Singh A, Safderi ZH, Kumar R. Allergic contact dermatitis by *Semecarpus Anacardium* for evil eye: A prospective study from Central India. *Indian Journal of Basic and Applied Medical Research* 3(3), 122-127, 2014.

[120] Vijayalakshmi T, Muthulakshmi V, Sachdanandam P. Toxic studies on biochemical parameters carried out in rats with Serankottai nei, a siddha drug-milk extract of *Semecarpus anacardium* nut. *J Ethnopharmacol* 69(1), 9-15, 2000.

[121] Lingaraju GM, Krishna V, Hoskeri HJ, Pradeepa K, Venkatesh, Babu PS. Wound healing promoting activity of stem bark extract of *Semecarpus anacardium* using rats. *Nat Prod Res* 26(24), 2344-2347, 2012.

[122] Tripathi YB, Pandey RS. *Semecarpus anacardium* L, nuts inhibit lipopolysaccharide induced NO production in rat macrophages along with its hypolipidemic property. *Indian J Exp Biol* 42(4), 432-436, 2004.

[123] Shukla SD, Jain S, Sharma K, Bhatnagar M. Stress induced neuron degeneration and protective effects of *Semecarpus anacardium* Linn. and *Withania somnifera* Dunn. in hippocampus of albino rats: an ultrastructural study. *Indian J Exp Biol* 38(10), 1007-1013, 2000.

[124] Vijayalakshmi T, Muthulakshmi V, Sachdanandam P. Effect of the milk extract of *Semecarpus anacardium* nut on adjuvant arthritis--a dose-dependent study in Wistar albino rats. *Gen Pharmacol* 27(7), 1223-1226, 1996.

[125] Nada R, Datta U, Deodhar SD, Sehgal S. Neutrophil functions in rheumatoid arthritis. *Indian J Pathol Microbiol* 42(3), 283-289, 1999.

[126] Narayan JP, John MS, Ghosh PK, Singh JN, Jha OP, Jha IS. Screening of some medicinal plants for spermatostatic and spermicidal properties. *Proceedings of Symposium on Phytochemistry and Botanical Classification*, CBS publishers and Distributors Pvt. Ltd, Delhi, India, 1985, 211-216.

[127] Asdaq SB, Prasannakumar SR. Protective effects of *Semecarpus anacardium* fruit extract against myocardial ischemia-reperfusion

injury in rats. *The Internet Journal of Alternative Medicine* 7(1), 2009.

[128] Tripathi YB, Pandey N, Tripathi D, Tripathi P. Oily fraction of *Semecarpus anacardium* Linn nuts involves protein kinase C activation for its pro-inflammatory response. *Indian J Exp Biol* 48(12), 1204-1209, 2010.

[129] Vijayakumar N, Subramanian P. Protective effect of *Semecarpus anacardium* against hyperammonia in rats. *Journal of Herbal Medicine and Toxicology* 2(4), 77-82, 2010.

[130] Premalatha B, Sachdanandam P. Potency of *Semecarpus anacardium* Linn. nut milk extract against aflatoxin B(1)-induced hepatocarcinogenesis: reflection on microsomal biotransformation enzymes. *Pharmacol Res* 42(2), 161-166, 2000.

[131] Rao SJM, Padmavathi V, Rao BK, Motohashi N. Cardenolides and relates of mainly *Calotropis Gigantea* and *C. Procera* in the Family Asclepiadacea. (Chapter 4) 109-180, 2015. In: Noboru Motohashi (ed.) *Occurrences, Structure, Biosynthesis, and Health Benefits Based on Their Evidences of Medicinal Phytochemicals in Vegetables and Fruits*. Volume 4. Nova Science Publishers, NY, USA, 2015.

[132] Padmavathi V, Rao BK, Motohashi N, Janardhan S, Sastry GN. Comparative and computer assisted drug designing of fatty acids isolated from flowers, leaves, stem bark, root bark and nuts of *Semecarpus anacardium* L.f. (Anacardiaceae). *J Pharm Pharmacol* 2, 582-591, 2014.

[133] Padmavathi V, Rao BK. Heterocyclic compound and their biological applications of *Semecarpus anacardium* L.f. Special chapter in the text book on *Heterocyclic Compounds*, 2017 (in press).

[134] Padmavathi V, Rao BK. Correlation between metal ions and organic compounds from *Semecarpus anacardum* l.f., their biological aspects and docking studies. *International Journal of Recent Scientific Research* 7(12), 14952-14959, 2016.

[135] Vustelamuri P, Bhattiprolu KR. Fatty acid composition of bhallataka oil and their biological properties. *Int J Pharm Bio Scis* 8(3), 81-92, 2017.

[136] (*Proceedings*) Padmavathi V, Rao KB. The potent molecules as drugs from *Semecarpus anacardium* L.f, Proceedings on FT-IR Studies in Designing and Discovering. [DDD-2013], Kolhapur, ISBN: 978-93-5126-349-4, RA-1, (pp 14-17), 2013.

[137] (*Proceedings*) Padmavathi V, Kesava R.B. New dimensions in isolation, anti-inflammatory, antimicrobial, multimetal analysis of *Semecarpus anacardium* L.f stem bark. *Proceedings of International Conference on New Dimensions in Chemistry and Chemical Technologies-Applications in Pharma Industry*, NDCST-2014, Jun 23-25, ISBN 978-93-82829-90-4, 422-427, 2014.

[138] (*Proceedings*) Padmavathi V, Rao BK. Naturally occurring biflavonoids from *Semecarous anacardium* L.f and their biological activities. *Proceedings of Andhra Pradesh Academi of Sciences (APAS)* 16(1), 39-43, 2014.

[139] (*Proceedings*) Padmavathi V, Rao KB. Proceedings on chemical constituents from methanolic extract of *Semecarpus anacardium* L.f. flowering buds, antimicrobial activity, anticancer activity and their docking studies. *Indian Youth Science Congress*, Jan 19-21, ANU Campus. This has got her Young Scientist Award in this Conference, 2015.

[140] (*Proceedings*) PadmavathiV, Kesava RB, Trend setting innovations of biflavonoids from *Semecarpus anacardium* L.f. of docking studies and biological activities in pharma industry. *Proceedings in International Conference on Trend Setting Innovations in Chemical Sciences and Technology-Applications in Pharma Industry* (TSCST-API, JNTUH), October 16-18, ISBN 978-93-82829- 48-5, pp.384-388, 2015.

[141] (*Proceedings*) Padmavathi V, Kesava RB. Physico chemical properties of *Semecarpus anacardium* L.f. seed oil. *Proceedings of International Conference on Trend Setting Innovations in Chemical*

Sciences and Technology, JNTUH, Hyderabad, October 4-6, OP 3, ISBN 978-93-82829-14-0, pp.224-231, 2016.

[142] (*Proceedings*) Padmavathi V, Kesava RB. Synthesis of epoxy *Semecarpus anacardium* L.f. oil and its evolution for lubricant properties. Poster presentation: 327, pp.360-361. *Proceedings of 104, Indian Science Congress. Tirupati, 2017.* Part-II, Session of chemical sciences, 3-7 Jan, 2017, held at Sri Venkateswara University, Tirupati.

In: Occurrences, Structure, Biosynthesis ... ISBN: 978-1-53614-141-2
Editor: Noboru Motohashi © 2018 Nova Science Publishers, Inc.

Chapter 4

CAROTENOIDS FROM MEXICAN PEPPERS AND THEIR BENEFICIAL EFFECTS

Carla Patricia Plazola-Jacinto[1,*],
Lourdes Valadez-Carmona[1,†],
D. Nayelli Villalón-López[2,‡]
and Marcela Hernández-Ortega[3,§]
[1]Departamento de Ingeniería Bioquímica,
Escuela Nacional de Ciencias Biológicas,
Instituto Politécnico Nacional, México
[2]Departamento de Química Orgánica,
Escuela Nacional de Ciencias Biológicas,
Instituto Politécnico Nacional, México
[3]Facultad de Ciencias de la Salud,
Universidad Anáhuac México Norte, México

---

* Email: patricia.plazola@gmail.com.
† Email: lvc24@hotmail.com.
‡ Email: lorienaule@gmail.com.
§ Corresponding Author Email: marcelahdz17@yahoo.com.mx.

ABSTRACT

Color in food is mainly related to the presence of different pigments, among which, carotenoids are the molecules responsible for the yellow, orange and red color of many fruits and vegetables. Peppers are fruits that possess and important amount of carotenoids, for this reason they are recognized as a good source of these phytochemicals. Of all carotenoids found in peppers, the capxanthin is the most characteristic pigment. Several studies have been focused on the beneficial effect of the carotenoids extracted from peppers, finding that these bioactive compounds could act as important antioxidants, protecting cells and tissues from harmful radical oxygen or nitrogen species. These pigments have also been related with analgesic and anti-inflammatory effects, being beneficial in many pathologies in which inflammation plays a key role. All the benefits provided by peppers carotenoids suggest that these fruits are an important alternative, not only to improve food taste and color, but also to help improve human health.

Keywords: pigments, peppers, carotenoids, *Capsicum annuum*

ABBREVIATIONS

ABA	abscisic acid
ASE	accelerated solvent extraction
Ar	argon
BTH	butylated hydroxytoluene
CD	circular dichroism
HPLC	high-performance liquid chromatography
IR	infrared
LC-PDA	liquid chromatography-photodiode array detection
MS	mass spectrometry
MAE	microwave-assisted extraction
NMR	nuclear magnetic resonance
N2	nitrogen
ORD	optical rotatory dispersion
PSY	phytoene synthase

PHWF	pressurized hot water extraction
PLE	pressurized liquid extraction
RP-TLC	reversed-phase thin-layer chromatography
NaCl	saline solution
SFE	supercritical fluid extraction
TBHQ	tertiary butylhydroquinone
THF	tetrahydrofuran
UAE	ultrasound-assisted extraction
UV	ultraviolet

1. INTRODUCTION

Food researches in carotenoids reside in the important antioxidants and provitamin A compounds, because humans are not able to synthesize *de novo* carotenoids, but are able to convert some provitamin A carotenoids into vitamin A by using intestinal monooxygenases [1, 2]. Researches on *Capsicum* genus pigment began in the 19th century. The paprika was the first pepper analyzed, and the carotenoids identified in it were β-carotene, cryptoxanthin, zeaxanthin, capsanthin, capsorubin, lutein epoxide, violaxanthin and mutatoxanthin. Since then, more than 700 natural carotenoids had been described and more than 20 new structures are reported each year [3, 4, 5].

The intake of dietary carotenoids has been associated with health benefits attributed to their antioxidant activity *in vivo, such as* the prevention and/or protection against cancer, heart disease and macular degeneration.

Capsicum species produce fruits that synthesize and accumulate carotenoid pigments in chromoplasts, which are responsible for the color yellow, orange and red in these fruits [4]. These pigments make them a potential raw material to industrial applications such as colorants mainly for human and animal food, feed natural additives to enhance the pigmentation mainly of fish and eggs, and their use in cosmetic and pharmaceutical products.

Currently the popularity of chili pepper fruits has been increasing, not only due its importance in human nutrition, but also due to its antioxidant properties and its beneficial effects. Pepper (*Capsicum annuum*) is a vegetable with wide number of varieties belonging to the Solanaceae family; its taxonomy includes two cultivars, the sweet ones, commonly eaten as vegetables and the hot ones, mostly used as a spice [6]. Chlorophylls and carotenoids are the most important pigments in pepper; they are located in the chloroplast and the chromoplast. It is proposed that carotenoids are accumulated in the center of the fibrils, and they are surrounded by a layer of polar lipids which, are surrounded by an outer layer of fibrillin that is directly attached to the plastid stroma in the chromoplast structure [4]. Xanthophyll epoxides are turned into ketoxanthophylls (capsanthin and capsorubin), responsible for the red color of ripe peppers, while chlorophyll provides green color and non-oxygenated carotenoids range from yellow to orange color (Figure 1) [6, 7].

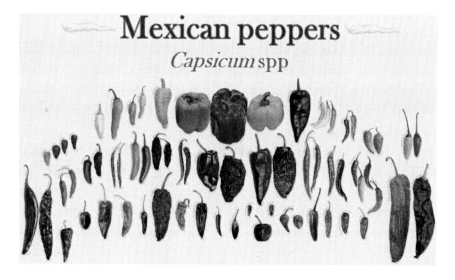

Figure 1. Mexican peppers varieties. Adapted from
http://www.biodiversidad.gob.mx/usos/alimentacion/chile.html.

2. CAROTENOID BIOSYNTHESIS IN CHILI PEPPER FRUITS

Carotenoids are 40-carbon isoprenoids with polyene chains that may contain up to 15 conjugated double bonds widely distributed in nature, which are synthesized by photosynthetic organisms such as cyanobacteria, algae and plants as well as non-photosynthetic microorganisms, such as fungi and some bacteria [4, 5, 8, 9]. In plants, the function of carotenoids involve cells and organelles protection against oxidative and light damage, making them a photoprotector by quenching chlorophyll triplet state of the photosynthetic apparatus, while xanthophyll epoxides serve as precursors for the synthesis of the abscisic acid hormone (ABA) (Figure 2) [3, 7, 8, 10].

In pepper, three independent pairs of genes (loci): *c1*, *c2*, and *y*, are known to control fruit color, and their allelic combinations yield eight different colors ranging from white (*y c1 c2*) to red (*y + c1 + c2+*). Camara [11] demonstrated that β-carotene, xanthophylls and keto-xanthophylls which have a β-epoxy-cyclohexenyl group are synthesized in the same way.

Pepper biosynthetic gene expression is related to high levels of total carotenoid accumulation; however its carotenoid biosynthesis regulation is not well understood. The biosynthetic steps for the *Capsicum* carotenoids have in common some genes and enzymes for the early steps of isoprenoids. The first step exclusive to carotenoid biosynthesis is the condensation of two molecules of the C-20 geranylgeranyl diphosphate by the phytoene synthase (PSY) to form phytoene, the backbone of carotenoids. Then, through a series of enzymatic desaturation, cyclization, and isomerization processes, phytoene is converted into a variety of carotenes and xanthophylls present in *Capsicum* [12, 13].

After phytoene formation, there are four sequential desaturations that lead to phytofluene, ζ-carotene, neurosporene and lycopene formation. Lycopene can subsequently undergo a cyclization by β-cyclase enzyme forming monocyclic compounds (e.g., γ-carotene, δ-carotene) and dicyclic compounds (e.g., α-carotene, β-carotene) [1, 4, 14]. Carotene undergoes through a β-ring hydroxylation on both rings generating zeaxanthin.

218 C.P. Plazola-Jacinto, L. Valadez-Carmona, D. Villalón-López et al.

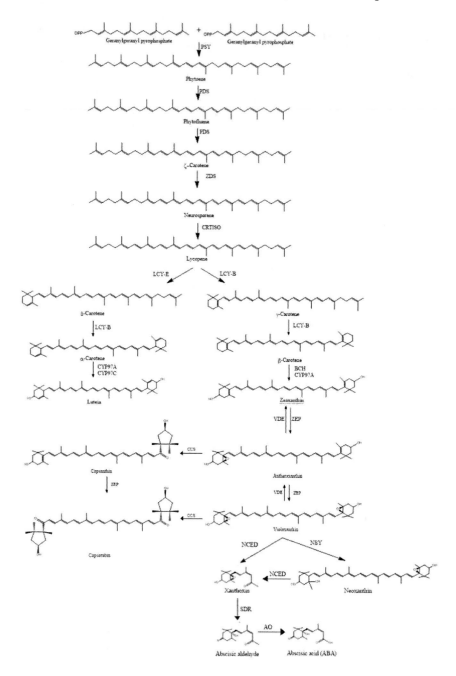

Figure 2. Biosynthesis of carotenoids.

Further epoxidation of both β-rings on zeaxanthin produces violaxanthin through the antheroxanthin intermediate. Violaxanthin can then be converted to neoxanthin by neoxanthins synthase. Both violaxanthin and neoxanthin can be isomerized and cleaved by 9-cis-epoxycarotenoid dioxygenases, providing xanthoxin, the apocarotenoid precursor to abscisic acid, and being this the final step of carotenoid biosynthesis [13].Carotenoid concentrations in chili peppers are determined mainly by two metabolic processes: 1) transformation of existing photosynthetic pigments; and 2) *de novo* carotenoid biosynthesis.

The pigment content varies during the several ripening stage and the chili pepper cultivars; for example, Bola and Agridulce peppers revealed that chlorophyllic pigments such as lutein and neoxanthin disappeared in the red I and red II stages, whereas β-carotene and violaxanthin (intermediates in the red pigment synthesis) increases their content while zeaxanthin, capsanthin, capsorubin, β-cryptoxanthin and capsolutein are synthesized *de novo*. Whereas, in yellow-orange chili peppers violaxanthin is the major carotenoid (37-68%), followed by *cis*-violaxanthin, antheraxanthin and lutein (5-14%).

In general, most of the carotenoids in ripe chili peppers are found in an esterified form with fatty acids, thus this provides them liposoluble characteristics and facilitates their accumulation in the lipophilic globule at the chromoplasts. The esterification process takes place mainly in *de novo* biosynthesized pigments such as capsanthin, capsorubin, zeaxanthin and β-cryptoxantin, sometimes esterification occurs in pigments synthesized previously such as violaxanthin. The red xanthophylls are esterified by short chains of saturated fatty acids, while yellow xanthophylls are esterified by myristic, palmitic, and unsaturated linoleic acid, making them less stables compared to the red ones due to the higher number of double bonds.

On the other hand, the full ripe chili peppers have three fractions, the free, the partially and the totally esterified forms of carotenoids, which make up the 21.3%, 35.6% and 43.1% of the total carotenoids respectively. In cultivars of full ripe chili peppers such as Numex, Mana, Negral, Belrubi and Delfin, lutein and neoxanthin, both characteristic chloroplast pigments, decreased in concentration with the ripening and eventually disappeared. β-Carotene, antheraxanthin, and violaxanthin increased in concentration, and other pigments were biosynthesized *de novo*: zeaxanthin, beta-cryptoxanthin, capsanthin, capsorubin, capsanthin-5,6-epoxide, and cucurbitaxanthin A, suggesting that changes in carotenoid content might be used as "ripeness index."

Capsanthin is the most synthesized and accumulated carotenoid in red *Capsicum spp.*, which represent up to 50% of the total carotenoid fruit content during the ripening stage [4]. Capsanthin is a powerful radical scavenger due to its structure conformed by 11 conjugated double bonds, a conjugated keto group and a cyclopentane ring.

The origin of the carotenoid color and properties is derived of the chromophore conjugated double bonds, which also is the responsible for their instability to oxygen, light and heat. For this reason, precautions have to be taken during analysis [15].

3. SAMPLE PREPARATION AND EXTRACTION FOR CAROTENOID ANALYSIS

The differences among the parts of the sample during carotenoid analysis have to be taken into account. As a consequence, the sample preparation as well as the solvent selection and extraction method is crucial, since its integrity can be affected by the destruction of the complexes with proteins or/and fatty acids. Many techniques propose the use of lyophilized raw material; some studies saponified the sample in order to hydrolyze carotenoid esters and to remove lipids and chlorophylls that may interfere with the carotenoids when using chromatographic detection [16, 17, 18, 19].

Currently, there are several techniques based on carotenoid spectroscopic and chromatographic properties, such as infrared (IR), nuclear magnetic resonance (NMR), mass spectrometry (MS), optical rotatory dispersion (ORD) and circular dichroism (CD), which is applied in stereochemistry [20, 21]. The general scheme to analyze carotenoids is shown in Figure 3.

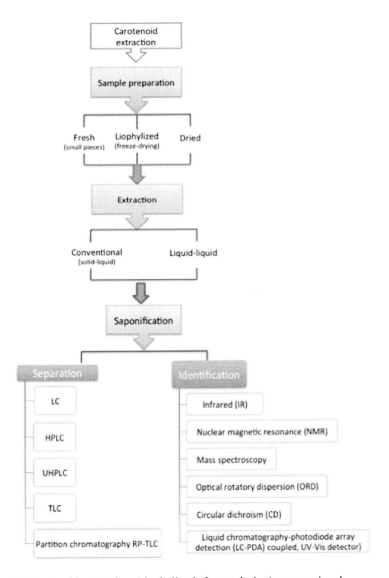

Figure 3. Carotenoid extraction. Liophylized (freeze-drying); conventional (solid-liquid); partion chromatography (reversed-phase thin-layer chromatography (RP-TLC)); nuclear magnetic resonance (NMR); mass spectroscopy; optical rotary dispersion (ORD); circular dichroism (CD), liquid chromatography-photodiode array detection (LC-PDA) coupled, UV-Vis detector).

As show the Figure 3, there are several sample preparation methods before carotenoid identification and quantification. The primary steps involve: extraction (using yellow/red light or absence of light), evaporation, saponification, and separation. All the steps have to be carry out in presence of protective nitrogen (N2) or argon (Ar) atmosphere at low temperatures). Usually antioxidants (e.g., butylated hydroxytoluene (BTH), pyrogallol, butylhydroxyanisol (BHA), ascorbic acid and tertiary butylhydroquinone (TBHQ), neutralizing agents (e.g., alkaline carbonates), are added to fresh pepper tissue and blanched in order to protect the sample from lipoxygenase activity [3, 16, 20, 22].

3.1. Extraction

To facilitate the carotenoid extraction, peppers must be ground or cut into small pieces, to avoid the oxidative or enzymatic degradation of carotenoids during the extraction, it should be carried out quickly with pure solvents, subjected to and ultrasonic force followed by mechanical shaking [23]. The most widely accepted solvents involve in liquid-liquid and solid-liquid extraction (usually for large volumes) are: dichloromethane, chloroform and tetrahydrofuran (THF), petroleum ether and hexane; either alone or in mixtures; also, the toxicity of the selected solvent should be considered. The polar compounds co-extracted with the carotenoids are traditionally removed by partitioning with aqueous salt solutions [3, 16, 17, 20, 24].

For a fresh pepper that contains 92% approximately of water, water-miscible polar organic solvents should be used such as acetone, methanol, ethanol or a solvent mixture, this is shown in Table 1; followed by lesser polar solvents to remove moisture. For dry pepper, a mixture of water-immiscible solvents may be used such as water-diethyl ether or petroleum ether with enough saturated saline solution (NaCl), in this case, the extraction is usually carried out simultaneously with the grinding in a blender and filtered (e.g.; by suction aid in a Buchner funnel or diatomaceous earth) [3, 16, 17, 20, 24].

Table 1. Analytical methods for detecting and quantification of carotenoid compounds in pepper samples

Method	Analytical instrument	Stationary phase	Mobile phase	Carotenoids isolated and/or identificated	Ref.
thin-layer chromatogram	TLC	silica gel 60 GF254	(A) hexane-EA-EtOH-acetone (95:3:2:2); (B) light petroleum ether (bp 65-95°C)- acetone-diethylamine (10:4:1); (C) methylene chloride-ethyl acetate (4:1)	7-9	[28]
		plates of silica gel 60 GF254 (20 × 20 cm glass plate, 0.7 mm thickness)	hexane/EA/EtOH/acetone (95:3:2:2)	5	[29]
		silica gel 60 GF254	hexane/EA/EtOH/acetone (95:3:2:2)	3	[30]
liquid-chromatogram	LC-DAD-MS	YMC analytical column, C-30 reversed-phase material, 5 µm (D_p = 4.6×250 mm), 35°C	MeOH, *tert*-butyl methyl ether, and water at 81:15:4, v/v/v (A) and 6:90:4, v/v/v (B), starting with 10 min isocratic at 100% A, gradient 50% to B at 40 min, 100% B at 50 min, 100% A at 55 min and isocratic 100% A at 55-60 min	26	[31]
	LC-DAD	ODS, D_p = 25 cm × 4.6 mm, 5 µm)	isocratic, mobile phase: ACN/MeOH/THF (58:35:7)	quantification of 3	[32]
	open column	MgO/hyflosupercel (1:1, activated for 4 hr, 110°C), packed to a height of 20 cm in D_p = 2.5 cm × 30 cm glass column	petroleum ether containing increasing amounts of ethyl ether and acetone	2	[25]

Table 1. (Continued)

Method	Analytical instrument	Stationary phase	Mobile phase	Carotenoids isolated and/or identified	Ref.
HPLC	HPLC-APCI-MS	YMC C-30 5 μm, $D_p = 4.6 \times 250$ mm	mixture of MeOH (A): MTBE (B) (50%, v/v) at 1 mL/min. gradient: t = 0 min 10% B; t = 2 min 10% B; t = 10 min 20% B; t = 20 min 70% B; t = 22 min 10% B; t = 25 min 10% B	9	[33]
	HPLC-DAD-APCI-TOF	C30, reversed-phase column, 3 μm, $D_p = 4.6 \times 250$ mm, 15°C	gradient elution of: water-MeOH-MTBE	53	[18]
	HPLC-DAD, HPLC-MS	C18 Spherisorb ODS2, 3 μm, $D_p = 4.6 \times 150$ mm	ACN (containing 0.05% of triethylamine)/ MeOH/ethyl acetate, used at a flow rate of 0.5 mL/min. (A) yellow pepper, a concave gradient; 95:5:0 to 60:20:20 in 20 min. (B) red pepper, concave gradient 99:1:0 to 60:20:20 in 20 min.	20	[25]
	HPLC-DAD-APCI-MS	YMC C30, 5 μm, $D_p = 4.6 \times 250$ mm	MeOH/MTBE/water (82:16:2, v/v/v; eluent A) and MeOH/MTBE/water, (10:88:2, v/v/v; eluent B), using a gradient program as follows: 0 min 0% B; 24 min 0% B; 150 min 100% B; 151 min 0% B	52	[34]
HPLC	HPLC-UV	YMC, C-30, column thermo-regulator, 5 μm, $D_p = 25 \times 4.6$ mm	gradient: acetone-water, flow rate 1.5 mL/min	6	[28]

Method	Analytical instrument	Stationary phase	Mobile phase	Carotenoids isolated and/or identificated	Ref.
HPLC	HPLC-UV	alkyl-silane modified silica C18, $D_p = 250 \times 4.6$ mm, 5 µm	THF-MeOH (10:90) or THF-CAN (isocratic (IS)):83 at a flow rate of 1.5 mL/min.	5	[24]
		C18, $D_p = 25$ cm × 4.6 mm; 5 µm	isocratic elution: ACN-MeOH-EA (73:20:7); 0.6 mL/min for 70 min, at 450 nm	6	[35]
		reversed-phase, ODS2-C18, $D_p = 250 \times 4.6$ mm, 5 µm	binary gradient: 1.5 mL/min, (A) acetone/water (75: 25, v/v) for 5 min. (B) linear gradient for 5 min 95: 5 (v/v).	3	[30]
		nucleodur, D_p = C18, 5 mm, 240 × 4.6 mm	0.01 M KH2PO4: MeOH: tetrabutylammonium hydroxide (97:3:0.1) at pH 2.8; 1 mL/ min	3	[36]
		normal-phase µ-porasil 125 Å, 10 µm, 3.9 × 150 mm reversed-phase C18 column 3.9 × 150 mm	gradient of hexane and acetone (9.5:0.5)	25	[37]
	HPLC-DAD	reversed-phase C18 Spherisorb ODS-2 (5 µm, $D_p = 0.46$ cm × 25 cm)	gradient: acetone-water, flow rate 1.5 mL/min	5	[29]
		reversed-phase C18 Spherisorb ODS 2, $D_p = 250 \times 4$ mm, 5 µm	gradient: acetone-water, internal standard: β-apo-8'-carotenal, flow rate 1.5 mL/min	20	[28]

Table 1. (Continued)

Method	Analytical instrument	Stationary phase	Mobile phase	Carotenoids isolated and/or identificated	Ref.
HPLC	HPLC-DAD	reversed-phase C18; $D_p = 250 \times 4.6$ mm, 5 µm	isocratic: ACN/MeOH/dichloro-methane (75:20:5, v/v/v); 10 mmol/L ammonium acetate, 4.5 mmol/L butylated hydroxytoluene (BHT), and 3.6 mmol/ L triethylamine at 1.5 mL/min	4	[38]
		reversed-phase, Inertsil ODS-3 C18, 5 µm, $D_p = 4.0 \times 250$ mm	(A) binary gradient acetone:water; 75:25 at 1 mL/min. (B) 95:5 for 5 min, washed with acetone	8	[39]
		nucleosil 100; $D_p = 250 \times 4.6$ mm, 3 µm	gradient: A: 9% water in MeOH, B: MeOH–ACN–isopropanol (10:35:55), and C: MeOH	-	[40]
	HPLC-DAD	nucleosil 120–3 µm column C18	isocratic elution with ACN/MeOH/ 2-propanol 85:10:5	---	[41]
		YMC C30, 5 µm, $D_p = 4.6 \times 250$ mm	A: MeOH/(MTBE)/H$_2$O (81:15:4) B: MTBE/MeOH (91:9); gradient elution was 100% A to 50% A and 50% B in 45 min; 100% B in 10 min, 100% A at 5 min; at 0.8 mL/min	9	[21]
		YMC C30, $D_p = 250 \times 4.6$ mm, 5 µm	MeOH/MTBE/water (83:15:2, v/v/v; A) and MeOH/MTBE/water (8:90:2); B); gradient: from 0-20 min 0%B; at 120 min 70% B; and at 125 min 100% B; 1.0 mL/min	Quantification 4	[42]

Method	Analytical instrument	Stationary phase	Mobile phase	Carotenoids isolated and/or identificated	Ref.
HPLC	HPLC	HPLC-DAD	25°C; 0.42 mL/min. MeOH/*tert*-butyl methyl ether/water (81:15:4, v/v/v; eluent (A) 4:92:4, v/v/v; eluent B). gradient: 0% B to 30% B (22 min), 30% B to 51.3% B (10 min), 51.3% B to 62.7% B (23 min), 62.7% B to 100% B (5 min), 100% B isocratic (5 min), 100% B to 0% B (5 min).	16	[43]
		Zorbax SB C18, 5 µm, $D_p = 3.0 \times 250$ mm	A: acetone-water (75:25, v/v); B: acetone-methanol (75:25, v/v) gradient: from 0 to 25% B in 10 min, from 25 to 100% B in 35 min, 100% B in 45 min and 0% B in 65 min. post time: 15 minutes	32	[44]
		nucleosil 100-3, C-18, Cyano, 5 µm, $D_p = 10 \times 250$ mm, analytical column	gradient A: 11% water in MeOH, B: MeOH and C: a mixture of isopropanol: ACN: MeOH 55:35:10 v/v/v %	43	[23]
		HPLC-DAD	gradient: MeOH/methyl-*tert*-butyl ether. (A) at 0.8 mL/min 80:20 to 0.5 min; (B) 75–15% at 0.5-15 min	-	[45]
		C30; $D_p = 4.6 \times 240$ mm, 3 µm	MTBE/MeOH/EtOAc (45:40:15)	4	[46]
UHPLC	NP-LC×RP-UHPLC/ PDA and LCMS–IT-TOF (APCI)	1D: one Ascentis ES Cyano column, $D_p = 250 \times 1.0$ mm, 5 µm 2D: LC × LC system, one Ascentis Express C18, $D_p = 30 \times 4.6$ mm, 2.7 µm d.p	1D mobile phase: (A) *n*-hexane; (B) *n*-hexane/butylacetate/acetone (80:15:5, v/v/v). 2D mobile phase: (A) water/ACN (10:90, v/v); (B) IPA.	33	[17]

Table 1. (Continued)

Method	Analytical instrument	Stationary phase	Mobile phase	Carotenoids isolated and/or identified	Ref.
supercritical fluid extraction (SFE)	UHPLC-DAD-APCI-AMMS	C18 SB, $D_p = 100 \times 2.1$ mm, 1.8 µm	gradient: 50:22.5:22.5:5 (v/v/v/v) water + 5 mM ammonium acetate/MeOH/ACN/EA at 35 °C (A) and 50:50 (v/v) ACN/EA (B), gradient was set as: (min/% A): 0.0/90, 0.1/90, 0.8/70, 20.0/9, 20.1/0, 20.4/0, 20.5/90, 23.0/90	12	[47]
	SC-CO$_2$, subcritical propane extraction	$D_p = 0.75 \times 30$ mm, packed with wide bore borosilicate coated with 1 µm thick Supelcowax film	(A) CO$_2$ at 35°C and 250 bar (B) solvent mixture rich in propane at 28°C and 100 bar, and 1.1 with propane at 25°C and 50 bar	recovered 4	[48]
	SC-CO$_2$	rushed-pellet particle size ($D_p = 0.273$–3.90 mm)	superficial solvent velocity $U_s = 0.57$–1.25 mm/s, and extraction pressure 320–540 bar, constant extraction temperature of 40°C	recovered 7	[49]
	SC-CO$_2$–NIR-Vis	C30, $D_p = 250 \times 4.6$ mm, with detector UV	60 to 80°C and 30 to 50 MPa using a CO$_2$/paprika mass ratio of 80:1; mass flow of 10 kg/h of CO$_2$	average carotenoid concentration β-carotene 26.8; free xanthophylls 24.2; xanthophyll and isomers esters 49.0 (% w/w)	[50]
	SC-CO$_2$	length 10.16 cm, 0.47 cm i.d., and 1.47 cm o.d.	cosolvents such as 1% EtOH or acetone, extraction pressure 140-480 bar at 40°C	Recovered 5	[51]

EA: ethyl acetate.

Conventional extraction is based on using solvents with low polarity such as hexane; however due to the carotenoids polar part, increasing the range of organic solvents is very useful. The use of high temperatures in solvents involve a small surface tension and viscosity, compared to the use of lower temperature, thereby, the solvent can reach easily the solute in certain matrix spaces, solubilizing several carotenoids. On the other hand, the solvent recovery has major disadvantages, which are the possible thermal degradation, the incomplete solvent removal, low selectivity and reproducibility [3, 16, 17].

In general, there is no specific method for pepper carotenoid extraction or a rule to select the ideal solvent. The influence of extraction time and the number of extractions, as well as, the application or not of a single solvent or a mixture, even the synergistic effect of solvents on extraction efficiency should be considered [16, 20].

Other traditional extraction techniques such as Soxhlet extraction, homogenization and sonication are often replaced by microwave-assisted extraction (MAE), supercritical fluid extraction (SFE) techniques, pressurized liquid extraction (PLE), accelerated solvent extraction (ASE), pressurized hot water extraction (PHWF) or ultrasound-assisted extraction (UAE). MAE shortens extraction time, is environmentally friendly, reduces solvent consumption and improves extraction yield depending on the dielectric constant of the extraction media [3, 16, 20].

3.1.1. Saponification

The saponification step do not affect carotenoids which are generally alkali-stable, and it is a purification procedure that removes unwanted lipids and chlorophylls; but during the procedure in certain conditions a carotenoid loss and isomerization can occur. Saponification is frequently employed prior to chromatographic analysis to simplify the carotenoid profile [3, 16, 17, 18, 21, 23, 25, 26].

The saponification procedure is usually carried out by adding 5-10% KOH in ethanol or methanol; other alternatives methods are boiling the alcoholic lipid solution for 30 minutes, leaving overnight at room temperature (in dark room) under nitrogen (N_2) atmosphere; however, to reduce the carotenoid loss, it has been suggested the use of low temperatures, inert atmosphere and presence of antioxidants [3, 16, 18, 21, 25].

Hydroxycarotenoids in red pepper are predominately esterified with fatty acids, hence the saponified extract is washed free of alkali, evaporated and dried in vacuum, the traces of water may be removed by adding sodium sulfate or/with azeotropic distillation with absolute ethanol [3, 16, 21, 25, 26].

4. SEPARATION, IDENTIFICATION AND QUANTIFICATION ANALYSIS OF CAROTENOIDS IN PEPPER

To date, both liquid chromatographic and spectrosphotometric methods have been extensively applied for carotenoids analysis; in particular high-performance liquid chromatography (HPLC) is the most popular as shown in Table 1, due to its low detection limit and its high reproducibility. However, it has many disadvantages, such as long time sample preparation and solvent extraction destroying their native complexes with proteins or fatty acids, the use of light that causes conformational change (structure) or carotenoid oxidation [27]. Nevertheless, it is already known that most of the pigments can absorb light in a particular ultraviolet (UV) spectrum due to a chromophore group, which carotenoids are characterized by conjugated double bonds responsible for light absorption and visible colors spectrum ranging from yellow to red [20].

Also, several LC-based techniques have been used where most carotenoids can be coupled to a photodiode array (PDA or DAD) or UV-Vis detector. Although LC separation coupled to UV-Vis instruments has been the most common analytical method for determining carotenoids qualitatively and quantitatively, the spectra of many carotenoids are very similar, so many researchers have complemented the identification of carotenoids using MS as detector [16].

For a brief summary of the existing methods to date for carotenoid analysis in pepper samples see Table 1; it is worth to mention that currently there is not a specific method or methodology for pepper carotenoid identification or pepper carotenoid isolation.

5. HEALTH BENEFITS RELATED TO PEPPER CAROTENOIDS

Peppers (*Capsicum* sp.) are worldwide consumed due to its bright colors and characteristic flavours. They are primarily used as fresh vegetables, spices or food colorants.

Besides the food uses, peppers have been widely recognized for their beneficial effects on human health. They possess chemopreventive characteristics, analgesic and anti-inflammatory activities [32, 52], antiobesogenic functions [53], as well as radical scavenging activity [54, 55].

These therapeutic applications are related to the presence of different phytochemicals such as capsaicinoids, phenolic compounds, and carotenoids [56, 57].

In regard to carotenoids, different epidemiological studies have shown the association between the consumption of carotenoid rich foods, such as peppers, and the reduction in several chronic diseases [58, 59].

Carotenoids possess diverse bioactive effects, which include provitamin A activity, immune modulation, antioxidant actions and involvement in cell signaling, being the absorption an important factor in the determination of the potential biological effects of carotenoids.

Studies of the carotenoid bioavailability from peppers found that, in these vegetables, color influence the carotenoid profile, bioaccessibility, and bioavailability of carotenoids rather than pepper type. As a result, it was shown that xanthophylls showed greater bioaccessibility than carotenes, suggesting that red peppers had a higher bioaccessibility than either the orange peppers or yellow peppers [38, 60, 61, 62]. Regard of the impact of the cooking method, it has been found that freezing and low temperature cooking methods improve the carotenoids bioavailability in peppers [62].

Safety assays of carotenoids extracted from peppers have proved that these pigments are safe to consume in doses even higher than 2,000 mg/kg [32]. Other studies have suggested that carotenoids from peppers do not generate genotoxicity, have antimutagenic effect against hydrogen peroxide (H_2O_2)-induced damage and significantly extend the lifespan and healthspan of *Drosophila melanogaster* at 1.25 and 2.5 mg/mL [63].

As chemopreventive agents, carotenoids isolated from red paprika (*Capsicum annuum* L.) showed potent *in vitro* antitumor activity. In the *in vivo* assay, the major carotenoids in paprika (capsanthin, capsanthin 3-ester and capsanthin 3,3'-diester) exhibited potent antitumor activity in a mouse skin two-stage carcinogenesis assay [64]. Fernández-Bedmar and Alonso-Moraga [63] found that pungent or non-pungent peppers, and capsanthin, principally red pigment in peppers, are able to inhibit the *in vitro* growth of human leukemia cell line HL60.

Peppers possess a wide range of carotenoids among which, capsanthin is the most representative red pigment. It has been studied for its capacity to ameliorate the impaired lipid metabolism in the liver and adipose tissue of high-fat diet-induced-obese mice due to the fact that it reduces weight gain and also improve serum lipid profile and adipokine secretion. Capsanthin consumption decreased lipid droplet (LD) size in adipose tissue and inhibited adipogenesis. In liver, the red pigment ameliorated hepatic steatosis, and this could be related to the suppressing effect on hepatic lipogenesis, fatty acid oxidation and gluconeogenesis. These findings suggest that capsanthin could be used as an alternative to decrease the detrimental effects of diet-induce obesity [65].

Diverse pathologies have as a pivotal component the inflammation and the pain related to it. Carotenoids from peppers have important analgesic and anti-inflammatory effects. A study performed by Hernández-Ortega, Ortiz-Moreno [32] found that carotenoids extracted from Mexican peppers exhibited significant peripheral analgesic activity at 5, 20, and 80 mg/kg and induced central analgesia at 80 mg/kg. These pigments significantly inhibited edema formation and progression at a dose of 5 mg/kg. The results obtained suggest that these pigments could be useful for pain and inflammation relief. The mechanism of action by which these pigments could exert the analgesic, and anti-inflammatory effects, is not yet clear, but may be related to their capacity to inhibit cyclooxygenase (COX) or other parts of the inflammation pathway.

Kim, Lee [66] suggests that carotenoids from peppers can protect WB-F344 rat liver epithelial cells from H_2O_2-induced inhibition of gap junction intracellular communication (GJIC) by recovering the mRNA expression encoding connexin 43 (C x 43) and preventing the phosphorylation of C x 43 protein (gap junction alpha-1 protein. GJA1). The red pigments of peppers significantly suppressed the formation of reactive oxygen species (ROS) in H_2O_2-treated cells. The results suggest that some of the beneficial effects exerted by carotenoids may be explained by the antioxidant effect that these pigments possess.

CONCLUSION

Peppers possess different types of bioactive molecules. Carotenoids are one of the most important phytochemicals which, not only generate good sensory attributes to food, but also has beneficial health related activities, suggesting that these carotenoid pigments are very important molecules in methods of improving human well-being.

REFERENCES

[1] Guzman I, Hamby S, Romero J, Bosland PW, O'Connell MA. Variability of carotenoid biosynthesis in orange colored *Capsicum* spp. *Plant Sci* 179(1-2), 49-59, 2010.

[2] Othman R, Mohd Zaifuddin FA, Hassan NM. Carotenoid biosynthesis regulatory mechanisms in plants. *J Oleo Sci* 63(8), 753-760, 2014.

[3] Arimboor R, Natarajan RB, Menon KR, Chandrasekhar LP, Moorkoth V. Red pepper (*Capsicum annuum*) carotenoids as a source of natural food colors: analysis and stability-a review. *J Food Sci Technol* 52(3), 1258-1271, 2015.

[4] Gómez-García Mdel R, Ochoa-Alejo N. Biochemistry and molecular biology of carotenoid biosynthesis in chili peppers (*Capsicum* spp.). *Int J Mol Sci* 14(9), 19025-19053, 2013.

[5] Hornero-Méndez D, Britton G. Involvement of NADPH in the cyclization reaction of carotenoid biosynthesis. *FEBS Lett* 515(1-3), 133-136, 2002.

[6] Conforti F, Statti GA, Menichini F. Chemical and biological variability of hot pepper fruits (*Capsicum annuum* var. acuminatum L.) in relation to maturity stage. *Food Chemistry* 102(4), 1096-1104, 2007.

[7] Bouvier F, d'Harlingue A, Hugueney P, Marin E, Marion-Poll A, Camara B. Xanthophyll biosynthesis. Cloning, expression, functional reconstitution, and regulation of beta-cyclohexenyl carotenoid epoxidase from pepper (*Capsicum annuum*). *J Biol Chem* 271(46), 28861-28867, 1996.

[8] Hirschberg J. Carotenoid biosynthesis in flowering plants. *Curr Opin Plant Biol* 4(3), 210-218, 2001.

[9] Nisar N, Li L, Lu S, Khin NC, Pogson BJ. Carotenoid metabolism in plants. *Mol Plant* 8(1), 68-82, 2015.

[10] Baranski R, Baranska M, Schulz H. Changes in carotenoid content and distribution in living plant tissue can be observed and mapped in situ using NIR-FT-Raman spectroscopy. *Planta* 222(3), 448-457, 2005.

[11] Camara B. Biosynthesis of keto-carotenoids in *Capsicum annum* fruits. *FEBS Letters* 118(2), 315-318, 1980.

[12] Kim OR, Cho MC, Kim BD, Huh JH. A splicing mutation in the gene encoding phytoene synthase causes orange coloration in Habanero pepper fruits. *Mol Cells* 30(6), 569-574, 2010.

[13] McQuinn RP, Giovannoni JJ, Pogson BJ. More than meets the eye: from carotenoid biosynthesis, to new insights into apocarotenoid signaling. *Curr Opin Plant Biol* 27, 172-179, 2015.

[14] Sandmann G. Carotenoid biosynthesis and biotechnological application. *Arch Biochem Biophys* 385(1), 4-12, 2001.

[15] Meléndez-Martínez AJ, Vicario IM, Heredia FJ. Review: Analysis of carotenoids in orange juice. *Journal of Food Composition and Analysis* 20(7), 638-649, 2007.

[16] Amorim-Carrilho KT, Cepeda A, Fente C, Regal P. Review of methods for analysis of carotenoids. *TrAC Trends in Analytical Chemistry* 56, 49-73, 2014.

[17] Cacciola F, Donato P, Giuffrida D, Torre G, Dugo P, Mondello L. Ultra high pressure in the second dimension of a comprehensive two-dimensional liquid chromatographic system for carotenoid separation in red chili peppers. *J Chromatogr A* 1255, 244-251, 2012.

[18] Cervantes-Paz B, Yahia EM, de Jesús Ornelas-Paz J, Victoria-Campos CI, Ibarra-Junquera V, Pérez-Martínez JD, Escalante-Minakata P. Antioxidant activity and content of chlorophylls and carotenoids in raw and heat-processed Jalapeño peppers at intermediate stages of ripening. *Food Chem* 146, 188-196, 2014.

[19] Díaz-Reinoso B, Moure A, Domínguez H, Parajó JC. Supercritical CO_2 extraction and purification of compounds with antioxidant activity. *J Agric Food Chem* 54(7), 2441-2469, 2006.

[20] Gross J (ed). *Pigments in vegetables: chlorophylls and carotenoids.* Carotenoids. pp 75-278, Springer; 1991.

[21] Pérez-López AJ, López-Nicolas JM, Núñez-Delicado E, Del Amor FM, Carbonell-Barrachina AA. Effects of agricultural practices on color, carotenoids composition, and minerals contents of sweet peppers, cv. Almuden. *J Agric Food Chem* 55(20), 8158-8164, 2007.

[22] Aman R, Carle R, Conrad J, Beifuss U, Schieber A. Isolation of carotenoids from plant materials and dietary supplements by high-

speed counter-current chromatography. *J Chromatogr A* 1074(1-2), 99-105, 2005.

[23] Nagy Z, Daood H, Koncsek A, Molnár H, Helyes L. The simultaneous determination of capsaicinoids, tocopherols, and carotenoids in pungent pepper powder. *Journal of Liquid Chromatography & Related Technologies* 40(4), 199-209, 2017.

[24] Epler KS, Sander LC, Ziegler RG, Wise SA, Craft NE. Evaluation of reversed-phase liquid chromatographic columns for recovery and selectivity of selected carotenoids. *J Chromatogr* 595(1-2), 89-101, 1992.

[25] de Azevedo-Meleiro CH, Rodriguez-Amaya DB. Qualitative and quantitative differences in the carotenoid composition of yellow and red peppers determined by HPLC-DAD-MS. *J Sep Sci* 32(21), 3652-3658, 2009.

[26] Kim S, Park J, Hwang IK. Composition of main carotenoids in Korean red pepper (*Capsicum annuum*, L) and changes of pigment stability during the drying and storage process. *Journal of Food Science* 69(1), FCT39-FCT44, 2004.

[27] Breithaupt DE. Simultaneous HPLC determination of carotenoids used as food coloring additives: applicability of accelerated solvent extraction. *Food Chemistry* 86(3), 449-456, 2004.

[28] Minguez-Mosquera MI, Hornero-Mendez D. Separation and quantification of the carotenoid pigments in red peppers (*Capsicum annuum* L.), paprika, and oleoresin by reversed-phase HPLC. *J Agric Food Chem* 41(10), 1616-1620, 1993.

[29] Hornero-Méndez D, Mínguez-Mosquera MI. Rapid spectrophotometric determination of red and yellow isochromic carotenoid fractions in paprika and red pepper oleoresins. *J Agric Food Chem* 49(8), 3584-3588, 2001.

[30] Fernández-García E, Carvajal-Lérida I, Pérez-Gálvez A. Carotenoids exclusively synthesized in red pepper (capsanthin and capsorubin) protect human dermal fibroblasts against UVB induced DNA damage. *Photochem Photobiol Sci* 15(9), 1204-1211, 2016.

[31] Breithaupt DE, Schwack W. Determination of free and bound carotenoids in paprika (*Capsicum annuum* L.) by LC/MS. *European Food Research and Technology* 211(1), 52-55, 2000.

[32] Hernández-Ortega M, Ortiz-Moreno A, Hernández-Navarro MD, Chamorro-Cevallos G, Dorantes-Alvarez L, Necoechea-Mondragón H. Antioxidant, antinociceptive, and anti-inflammatory effects of carotenoids extracted from dried pepper (*Capsicum annuum* L.). *J Biomed Biotechnol* 2012;2012:524019. doi: 10.1155/2012/524019. Epub 2012 Oct 2.

[33] Guil-Guerrero JL, Martínez-Guirado C, del Mar Rebolloso-Fuentes M, Carrique-Pérez A. Nutrient composition and antioxidant activity of 10 pepper (*Capsicum annuun*) varieties. *European Food Research and Technology* 224(1), 1-9, 2006.

[34] Giuffrida D, Dugo P, Torre G, Bignardi C, Cavazza A, Corradini C, Dugo G. Characterization of 12 *Capsicum* varieties by evaluation of their carotenoid profile and pungency determination. *Food Chem* 140(4), 794-802, 2013.

[35] Troconis-Torres IG, Rojas-López M, Hernández-Rodríguez C, Villa-Tanaca L, Maldonado-Mendoza IE, Dorantes-Álvarez L, Tellez-Medina D, Jaramillo-Flores ME. Biochemical and molecular analysis of some commercial samples of chili peppers from Mexico. *J Biomed Biotechnol* 2012;2012:873090. doi: 10.1155/2012/873090. Epub 2012 May 14.

[36] Koncsek A, Kruppai L, Helyes L, Bori Z, Daood HG. Storage stability of carotenoids in paprika from conventional, organic and frost-damaged spice red peppers as influenced by illumination and antioxidant supplementation. *Journal of Food Processing and Preservation* 40(3), 453-462, 2016.

[37] Collera-Zúñiga O, Jiménez FG, Gordillo RM. Comparative study of carotenoid composition in three Mexican varieties of *Capsicum annuum* L. *Food Chemistry* 90(1-2), 109-114, 2005.

[38] O'Sullivan L, Jiwan MA, Daly T, O'Brien NM, Aherne SA. Bioaccessibility, uptake, and transport of carotenoids from peppers (*Capsicum* spp.) using the coupled *in vitro* digestion and human

intestinal Caco-2 cell model. *J Agric Food Chem* 58(9), 5374-5379, 2010.

[39] Topuz A, Dincer C, Özdemir KS, Feng H, Kushad M. Influence of different drying methods on carotenoids and capsaicinoids of paprika (Cv., Jalapeno). *Food Chemistry* 129(3), 860-865, 2011.

[40] Daood HG, Biacs PA. Simultaneous determination of Sudan dyes and carotenoids in red pepper and tomato products by HPLC. *J Chromatogr Sci* 43(9), 461-465, 2005.

[41] Simkin AJ, Zhu C, Kuntz M, Sandmann G. Light-dark regulation of carotenoid biosynthesis in pepper (*Capsicum annuum*) leaves. *J Plant Physiol* 160(5), 439-443, 2003.

[42] Giuffrida D, Dugo P, Torre G, Bignardi C, Cavazza A, Corradini C, Dugo G. Evaluation of carotenoid and capsaicinoid contents in powder of red chili peppers during one year of storage. *Food Research International* 65(Part B), 163-170, 2014.

[43] Schweiggert U, Kurz C, Schieber A, Carle R. Effects of processing and storage on the stability of free and esterified carotenoids of red peppers (*Capsicum annuum* L.) and hot chili peppers (*Capsicum frutescens* L.). *European Food Research and Technology* 225(2), 261-270, 2007.

[44] Kevrešan ŽS, Mandić AP, Kuhajda KN, Sakač MB. Carotenoid content in fresh and dry pepper (*Capsicum annuum* L.): Fruits for paprika production. *Food Processing, Quality and Safety* 36(1-2), 21-27, 2009.

[45] de Lima Petito N, da Silva Dias D, Costa VG, Falcão DQ, de Lima Araujo KG. Increasing solubility of red bell pepper carotenoids by complexation with 2-hydroxypropyl-β-cyclodextrin. *Food Chem* 208, 124-131, 2016.

[46] Ishida BK, Chapman MH. Carotenoid extraction from plants using a novel, environmentally friendly solvent. *J Agric Food Chem* 57(3), 1051-1059, 2009.

[47] Bijttebier S, Zhani K, D'Hondt E, Noten B, Hermans N, Apers S, Voorspoels S. Generic characterization of apolar metabolites in red

chili peppers (*Capsicum frutescens* L.) by orbitrap mass spectrometry. *J Agric Food Chem* 62(20), 4812-4831, 2014.
[48] Illés V, Daood HG, Biacs PA, Gnayfeed MH, Mészáros B. Supercritical CO_2 and subcritical propane extraction of spice red pepper oil with special regard to carotenoid and tocopherol content. *Journal of Chromatographic Science* 37(9), 345-352, 1999.
[49] Uquiche E, del Valle JM, Ortiz J. Supercritical carbon dioxide extraction of red pepper (*Capsicum annuum* L.) oleoresin. *Journal of Food Engineering* 65(1), 55-66, 2004.
[50] Ambrogi A, Cardarelli DA, Eggers R. Fractional extraction of paprika using supercritical carbon dioxide and on-line determination of carotenoids. *Journal of Food Science* 67(9), 3236-3241, 2002.
[51] Jarén-Galán M, Nienaber U, Schwartz SJ. Paprika (*Capsicum annuum*) oleoresin extraction with supercritical carbon dioxide. *J Agric Food Chem* 47(9), 3558-3564, 1999.
[52] Chen L, Kang Y-H. Anti-inflammatory and antioxidant activities of red pepper (*Capsicum annuum* L.) stalk extracts: Comparison of pericarp and placenta extracts. *Journal of Functional Foods* 5(4), 1724-1731, 2013.
[53] Sung J, Bang MH, Lee J. Bioassay-guided isolation of anti-adipogenic compounds from defatted pepper (*Capsicum annuum* L.) seeds. *Journal of Functional Foods* 14, 670-675, 2015.
[54] Hisatomi E, Matsui M, Kubota K, Kobayashi A. Antioxidative activity in the pericarp and seed of Japanese pepper (*Xanthoxylum piperitum* DC). *J Agric Food Chem* 48(10), 4924-4928, 2000.
[55] Zimmer AR, Leonardi B, Miron D, Schapoval E, Oliveira JR, Gosmann G. Antioxidant and anti-inflammatory properties of *Capsicum baccatum*: from traditional use to scientific approach. *J Ethnopharmacol* 139(1), 228-233, 2012.
[56] De Marino S, Borbone N, Gala F, Zollo F, Fico G, Pagiotti R, Iorizzi M. New constituents of sweet *Capsicum annuum* L. fruits and evaluation of their biological activity. *J Agric Food Chem* 54(20), 7508-7516, 2006.

[57] Tundis R, Loizzo MR, Menichini F, Bonesi M, Conforti F, Statti G, De Luca D, de Cindio B, Menichini F. Comparative study on the chemical composition, antioxidant properties and hypoglycaemic activities of two *Capsicum annuum* L. cultivars (Acuminatum small and Cerasiferum). *Plant Foods Hum Nutr* 66(3), 261-269, 2011.

[58] Cooper DA. Carotenoids in health and disease: recent scientific evaluations, research recommendations and the consumer. *J Nutr* 134(1), 221S-224S, 2004.

[59] Rao AV, Rao LG. Carotenoids and human health. *Pharmacol Res* 55(3), 207-216, 2007.

[60] Granado-Lorencio F, Olmedilla-Alonso B, Herrero-Barbudo C, Blanco-Navarro I, Pérez-Sacristán B, Blázquez-García S. *In vitro* bioaccessibility of carotenoids and tocopherols from fruits and vegetables. *Food Chemistry* 102(3), 641-648, 2007.

[61] Maiani G, Castón MJ, Catasta G, Toti E, Cambrodón IG, Bysted A, Granado-Lorencio F, Olmedilla-Alonso B, Knuthsen P, Valoti M, Böhm V, Mayer-Miebach E, Behsnilian D, Schlemmer U. Carotenoids: actual knowledge on food sources, intakes, stability and bioavailability and their protective role in humans. *Mol Nutr Food Res* 53(Suppl 2), S194-S218, 2009.

[62] Pugliese A, O'Callaghan Y, Tundis R, Galvin K, Menichini F, O'Brien N, Loizzo MR. *In vitro* assessment of the bioaccessibility of carotenoids from sun-dried chili peppers. *Plant Foods Hum Nutr* 69(1), 8-17, 2014.

[63] Fernández-Bedmar Z, Alonso-Moraga A. *In vivo* and *in vitro* evaluation for nutraceutical purposes of capsaicin, capsanthin, lutein and four pepper varieties. *Food Chem Toxicol* 98(Pt B), 89-99, 2016.

[64] Maoka T, Mochida K, Kozuka M, Ito Y, Fujiwara Y, Hashimoto K, Enjo F, Ogata M, Nobukuni Y, Tokuda H, Nishino H. Cancer chemopreventive activity of carotenoids in the fruits of red paprika *Capsicum annuum* L. *Cancer Lett* 172(2), 103-109, 2001.

[65] Kim JS, Ha TY, Kim S, Lee SJ, Ahn J. Red paprika (*Capsicum annuum* L.) and its main carotenoid capsanthin ameliorate impaired

lipid metabolism in the liver and adipose tissue of high-fat diet-induced obese mice. *Journal of Functional Foods* 31, 131-140, 2017.

[66] Kim JS, Lee WM, Rhee HC, Kim S. Red paprika (*Capsicum annuum* L.) and its main carotenoids, capsanthin and β-carotene, prevent hydrogen peroxide-induced inhibition of gap-junction intercellular communication. *Chem Biol Interact* 254, 146-155, 2016.

ABOUT THE EDITOR

Noboru Motohashi, PhD

Foreign Examiner for Evaluation of the PhD Thesis in *Indian National University*, India; Former Professor and Director (elected Riji in Japanese) of *Meiji Pharmaceutical University*, Tokyo Japan.
Email: noborumotohashi@jcom.home.ne.jp

Professional Appointments:

- 1966: Pharmaceutical Sciences, *Tohoku University*, Sendai, Japan (graduated school)
- 1985-1986: Visiting Professor of Department of Medicinal Chemistry, College of Pharmacy, *University of Kansas* (KU), Lawrence, KS, USA
- 2006: Professor of *Meiji Pharmaceutical University*, Tokyo, Japan (retired)
- 2007: Advisor of *Meiji Pharmaceutical University*, Tokyo, Japan (retired)
- 2011: Elected Director (Riji in Japanese) of *Meiji Pharmaceutical University*, Tokyo, Japan (retired)

INDEX

#

1,8-cineole (32), 145
13-hydroxy-7-oxoazorellane (40), 153
16-O-methyl-cafestol, 108
1-pentadeca-7,10-dienyl-1,3-dihydroxybenzene (130), 181
1-pentadeca-8-enyl-2,3-dihydroxybenzene (131), 181, 183
2-amino-1-methyl-6-phenylimidazopyridine (PhIP), 124
3-((7E,10E)-pentaca-7,10-dienyl)benzene-1,2-diol (105), 181
3-caffeoylquinic acid, 105, 120, 121
3-pentadecyl benzene (136), 180, 181
3T3-L1 adipocytes, 35
5-O-caffeoylquinic acid (neochlorogenic acid. 17), 105, 118
7,31,41-trihydroxyflavone (140), 188
7-deacetylazorellanol (41), 153
7-methylxanthosine (19), 95, 106
9-cis-epoxycarotenoid dioxygenases, 217

A

A1 and A2A adenosine receptors, 123
acrid stimulant, 171
actoprotectors and adaptogens, 2
acute toxicities, 154
adenine nucleotide, 106
adenosine A1 receptor, 120
adenosine receptor antagonist (A1, A2A, A2B, and A3), 123
aflatoxin B1, 95, 124
aflatoxin B1 (AFB1)-DNA adducts, 124
aging, 48
agonists, 36
alanine aminotransferase (ALT), 122
alkaline carbonates, 220
alkaloid, 159, 193
amentoflavone (36), 152, 181
American ginseng, 56
anacardiaceae family, 134
anacardic acid (129), 181
anacardoflavonone (134), 181
anacardoside (135), 179, 181
analgesic activity, 230
analgesic effect, 153

Index

Anastatica hierochuntica, 152
anastatin B (38), 152
antagonist, 6
antheroxanthin intermediate, 217
anthocyanins, 97
anticarcinogenic, 123
anti-hepatocarcinogenesis effects, 124
anti-inflammatory, xi, 153
anti-inflammatory effect, xi
antimicrobial agent, 123, 148
antioxidant, 144, 191
antioxidant activity, 145
antioxidant capacities, 94
antioxidant replacements, 144
antioxidants, 144, 220
antiproliferative effect, 94
antispasmodic, 171
apocarotenoid precursor, 217
arabica, viii, 99, 112, 114, 116, 125, 127, 129, 131
arabinogalactans, 124
arabinose, 124
aroma, 97, 100, 109
aromatic and medicinal plants, 144
ascorbic acid, 220
aspartate aminotransferase (AST), 122
asthma, 153, 171, 172
asthma symptoms, 123
atherosclerotic lesion, 120
atherothrombosis, 120
auto-oxidation, 144
autumnal Ginkgo biloba leaves, 152
availability, 55
Azorella compacta, 153
Azorella yareta, 153
azorellanol (39), 153

B

B16-BL6 melanoma cells, 19
bacteria adsorption ability, 123
bacterial colon population, 124
bactericidal activity, 125, 191
bacteriostatic activity, 125
basal metabolic rate (BMR), 123
benzodiazepine receptors, 152
benzofuran moiety, 152
berries, 99
beverage, viii, 93, 97, 99, 117, 123, 124, 125
bhilwanol (128), 181
biflavones, 152
biflavonoids, 184
bioactive chemical constituents, 103
bioactive compounds, xi, 97, 103, 119, 122, 128, 129, 130, 139, 153, 212
bioactive molecules, 125, 231
bio-compounds, 110
biofilm, 124
biofilm formation, 124
biogenetic pathways, 156
biosynthesis, 152
blood, 10, 175, 196
blood glucose, 39, 40, 41
blood glucose levels, 40, 120, 192
blood pressure, vii, 10
blood pressure decrease, 10
blood pressure drop, 11
blood pressure lowering effect, 11
blood purifier, 175
blood-brain barrier (BBB), 106
body fat catabolism, 121
body resistance, 176
bones, 174
brain, 144, 175
brain health, 121
brain homogenates, 144
brain tonic, 175
brewed coffee, 119, 122
brewing methods, 122
bronchitis, 153
browning intensity (BI), 15
bruised nut, 172

Index

bruises, 174
butylated hydroxytoluene (BTH), 220
butylhydroxyanisol (BHA), 220

C

cafestol, 94, 97, 101, 103, 108, 110, 111, 119, 124, 127, 129
cafestol (CAF. 1), 94, 103, 108, 110, 111, 119, 124
caffeic acid (8), 94, 102, 103, 122
caffeinated coffee, 122
caffeine, 114, 116
caffeine (1,3,7-trimethylxanthine. 4), 94, 119, 125
caffeine biosynthesis, 106
caffeine biosynthetic pathway, 106
caffeine synthase (CS), 106
caffeoyl-CoA (16), 94, 105
caffeoyl-D-glucose (15), 94, 105
caffeoylquinic acid, 94, 97, 103, 118, 120, 121
caffeoylquinic acid (10), 94, 103
cancer, 192
caramelization, 109
carcinogen-induced liver cancer, 123
carcinogens, 124
carnitine, 95
carnitine palmitoyltransferase I (CPT1), 123
carotenoid biosynthesis, 217
carotenoid identification, 220, 229
carotenoids, 227
carvacrol (26), 145
catechins, 97
cell damage, 78
cell membranes, 144
cell signaling, 195
cells, 15, 18, 20, 21, 23, 33, 34, 35, 36, 38, 39, 42, 51, 53, 54, 55, 73, 78, 143, 191, 192, 195, 231
centrifugation, 119

chemical-induced neoplastic transformation, 120
chloroform, 116, 191
chlorogenic acid, 94, 95, 101, 103, 104, 105, 109, 113, 115, 117, 120, 122, 125, 127, 129
chlorogenic acid (CGA. 3-caffeoylquinic acid. 3), 94, 104, 125
chlorogenics acids, 97
cholagogue, 172
cholesterol, 41, 192
chromaffin cells, 49, 54
chronic liver disease risk, 122
chronic rheumatic disorders, 175
cinnamic acid, 103, 109
circulating forms, 122
Coffea arabica, 94, 99, 113, 115, 122, 127, 128, 129
Coffea canephora, 94, 99
coffee, 112, 114, 116
coffee brew, 94, 117, 122, 124, 130
coffee chemicals, 110
coffee cherries, 99
coffee consumption, 97, 120, 122, 126, 130
coffee crops, 93
coffee fruit, 99
coffee polyphenols, 121
cold pressing, 110
colds, 153
colonic microflora, 124
color, 144
colorants, 229
combination, 175
complementary DNAs (cDNAs), 152
compound K (CK), 23
confection, 171
consumption, 60, 61, 65, 69
copalyl diphosphate synthase, 108
copolymerization reactions, 124
corpus cavernou, 46
cough, 172
cryptoxanthin, 218

cytochrome P450 (CYP) isoenzymes 1A1, 1A2, 3A2, 2B1, 2B2, and 2C11, 124
cytochrome P450 isoform CYP1A2, 123

D

de novo route, 106
death, 15, 159
decaffeinated coffee, 119, 120
density, 14, 24, 47, 195
dental caries protection, 124
depression risk reduction, 123
D-galactosamine-induced cytotoxicity, 152
diabetes, 6, 176
diabetes of kapha-type, 176
dicaffeoylquinic acid (13), 94, 103, 105
differentiation, 13, 36
digestive, 171
disruption, 125
diterpene alcohols, 97
diterpenes, 94, 101, 109, 110, 111, 118, 119, 124, 125, 128, 129
diterpenoids, 154
dried root, 15
dysmenorrheal, 175
dyspepsia, 171, 172

E

edema, 231
Egyptian medicinal herb, 152
elution solvent, 118
endo-peptidases, 96, 124
endothelial leukocyte adhesion molecule-1 (E-selectin. ELAM-1), 120
endothelial protection, 120
energy intake, 121, 130
ent-copalyl diphosphate (ent-CPP), 108
ent-kaurene (20), 95, 108
ent-kaurene derivatives, 108
enzymatic activities, 120

enzymes, 152, 191, 195
epidemiological studie, 121, 229
epilepsy, 171, 172
epoxidation, 217
erectile dysfunction (E.D.) or impotence, viii, 2, 82, 83
erection, 45
escharotics, 171
ester bond, 123
esterase enzymes, 122
ethanol, 111, 116, 184
evaporation, 220
excitation, 11
exercise performance, 59
exothermic pyrolysis, 100
expression, 13, 14, 23, 174, 196
extract, 17, 33, 34, 35, 36, 37, 38, 39, 40, 41, 42, 43, 44, 45, 47, 55, 61, 80, 81, 82, 86, 90, 91, 110, 111, 113, 114, 115, 116, 117, 119, 127, 129, 142, 144, 152, 159, 182, 188, 190, 191, 192, 193, 194, 195, 196, 204, 205, 206, 207, 208, 209, 228
extraction, 5, 15, 96, 111, 220, 226, 227
extraction methods, 110, 111, 129
extraction processes, 110
extraction yield (EY), 15

F

fat, 12, 67
fat oxidation, 67, 123
fatigue, 48, 61, 88, 106
fatty acid, 5, 30, 36, 41, 58, 60, 64, 69, 95, 192, 195
fatty acid synthase (FAS), 123
ferric-ion-stimulated lipid peroxidation, 144
ferulic acid, 94, 97, 102, 103
ferulic acid (9), 94, 102, 103
fibroblast growth factor (FGF2)-induced neo-vascularization, 23
flavonoids, 112, 152

flavonoids,biflavonoids and pharmacological activities, 134
food, 144
food industry, 144
food lipids, 144
food quality, 144
formulation, 61, 65, 68, 69
free radical production, 144
free radicals, 144
fresh pepper tissue, 220
fresh raw *Panax ginseng* C.A. Meyer, 15
fruit, 171, 193, 195, 218
function, 10, 13, 64, 71, 72, 192, 195
functional beverage, 103
functionality, 144
furan group, 108

G

galactomannans, 124
galactose, 124
gamma-glutamyl transferase (GGT), 122
gastrointestinal tract, 23
genes, 152
geranylgeranyl diphosphate (GGPP), 108
Ginkgo biloba, 152
ginseng, 5, 15, 23
ginseng extractions, 16
ginseng marc, 5
ginseng root, 5, 19, 45, 87
ginseng saponins (ginsenosides), viii, 2, 48, 83
ginsenoside Rb1, 23
ginsenoside-Rb 2, 2
glucagon, 5
glucagon-like peptide-1 (GLP-1), 120
glucokinase (GK), 123
glucose, 5, 95
glucose (22), 95, 121, 123
glucose absorption, 120
glucose cyclization, 103

glucose metabolism regulator, 120
glucose uptake, 33, 120
glucose-6-phosphatase (G6Pase) activity, 120
glucose-dependent insulinotropic polypeptide (GIP), 120
glutamate-receptor gene GRIN2A, 122
glutathione peroxidase (GPx), 124
glutathione reductase (GR), 124
glutation S-transferase (GST), 120
gold wire frog, 10
green and roasted coffee beans, 108, 110
green coffee, 99, 109, 110, 111, 113, 115, 119, 128, 129
green coffee beans, 99, 110, 111, 113, 115, 119
growth cycle, 103
gut microbiota, 122
gut peptides, 120

H

habitual caffeinated-coffee consumption, 121
habitual coffee consumption, 97, 125
haematinic tonic, 175
heart, 10, 195
heart tissue, 123, 195, 196
heat, 174
hepatic cell damage, 122
hepatic gluconeogenesis, 120
hepatic glucose-6-phosphatase activity, 120
hepatocytes, 152
hepatoma HepG2 human cells, 124
hepatoprotective effects, 152
HepG2 cells, 37
herbs, 8, 48
hexane, 191, 227
high performance liquid chromatography (HPLC), 118
homocysteine (23), 95, 121

Index

human cancer cell lines, 94
human intestinal bacterial flora, 23
human plasma, 122
human umbilical vein endothelial cells (HUVEC), 23
humans, xi, 60, 83, 89, 121, 122, 213, 238
hydrophilic fraction, 117
hydroxycinnamic acid (21), 95, 121
hydroxycinnamic acids, 97
hydroxycinnamoyl-coenzyme A quinate transferases (HQT), 105
hydroxylases, 103
hydroxylation, 105, 215
hypericum perforatum, 152
hypoglycemic, viii, xi, 33, 44, 69, 81, 83, 123

I

I-41,II-31,41,I-5,II5,I7-Hexahydroxy[I3,II8]biflavonone (101), 181
I-41,II-41,I-5,II-5,I-7,II-7-hexahydroxy[I-3,II-8]biflavonone(31,8-binaringenin) (102), 181
illnesses with inflammation, 153
immune response, 196
impaired glucose tolerance, 120
India, 241
inflammation biomarkers, 122
inflammatory processes, 37
infusion, 81, 97
inhibition, 191
initial conversion, 106
innocuousness, 154
insulin, 5, 6
insulin functions, 120
insulin sensitizer, 120
intercellular adhesion molecule-1 (ICAM-1), 120
intercellular communication mediator, 23
intermediate, x, 6, 71, 72, 105, 182, 195, 196, 233
internal use, 171
intravenously (i.v.), 10
iodine vapor, 118
IR, 218
iron acquisition, 125
iron chelation, 125
ischemia, 195
ischemic heart disease, 97
isopentenyl diphosphate (IPP), 108
isoprenoids, 215

J

jaggery, 175
joints, 174
juice, 172, 174

K

kahweol, 94, 97, 101, 103, 108, 110, 111, 119, 124, 127, 129
kahweol (KAH. 2), 94, 103, 108, 110, 111, 119, 124
kaurene synthase (KS), 108
kidney, 192
Korean red ginseng, 5, 45, 87

L

lack- stamina, 2
lactate dehydrogenase (LDH) activity, 191
lactic acid dehydrogenase, 5
large bowel, 122
leaves, ix, 142, 156, 185
leprosy, 171
leprotic nodules, 174
leprous, 171, 172
leprous disease, 171

limonene (31), 145
linalool (29), 145
lipid, 33, 144, 239
lipid components, 144
lipid metabolism enzyme regulator, 123
lipids, 195
lipoxidation, 124
lipoxygenase activity, 220
liver, 192
liver cancer cell proliferation, 123
liver cell injury, 122
liver tissue, 123
long-term and habitual coffee consumption, 120
loss, 227, 228
low temperatures, 220
luciferase activity, 34, 35

methyl donor, 106
methyl tert-butyl ether (MTBE), 119
methylxanthines, 97, 106, 128
Mexican coffee, 117
microorganisms, 191
microtiter plates, 124
mild cognitive impairment subjects, 121
milk, 172, 191, 196
mobile phase, 118, 119, 221, 225
molecules, 191
mouth, 172
mRNA level, 36
mucilage, 99
multiplicity, 123
muscle, 10, 46
muscle spasm conditions, 10
myocardium, 10
myrcene (28), 145

M

maceration, 110
mannans, 124
Matrigel, 23
Matrigel plugs assay, 23
matrix metalloproteinase 1 (MMP-1), 124
matrix metalloproteinase 2 (MMP-2), 23
matrix metalloproteinase 9 (MMP-9), 23
matrix spaces, 227
MCF-7 cells, 191
mechanisms, 34, 125
medicated milk, 176
medium, 61, 112
melanoidin content, 124
melanoidins, 94, 97, 103, 109, 124, 131
menstruation, 175
metabolism, 230, 239
metal-chelating properties, 125
metalloproteinases, 23, 96
metastatic inhibition, 23
metformin (24), 95, 120, 121
methanol, 111, 113

N

n-cumaric acid, 97
neoxanthin, 217
neoxanthins synthase, 217
nervine, 171
nervous debility, 171, 172
nervous system, 4, 95, 143, 172
neuralgia, 172
neurodegenerative illness, 94
neuroprotective, viii, 2, 78, 83, 90, 123, 155, 194
neuroprotective effect, viii, 2, 78, 83, 90, 194
neutralizing agents, 220
nitric oxide, 196
nitric oxide bioavailability, 120
nitrogenous compounds, 97
N-methyl nucleosidase, 106
N-methyltransferases, 106
N-nitrosodimethylamine (NDMA), 124

non-esterified fatty acids (NEFAs. free fatty acids), 64, 69
normal-weight subjects, 123
nuclear export, 23
nuclear factor (NF)-kappaB-p65, 23
nuclear respiratory factor 2 (Nrf2), 123
nut, 171, 172, 174, 184, 191, 193
nutritive value, 144

O

odor, 109, 125
oil, 144, 172, 174, 176
oligomenorrhea, 176
oral administration, 12, 13, 14, 23
organic phase, 119
organoleptic characteristics, 97
organoleptic features, 125
organoleptic qualities, viii, 94
outer membrane, 37, 68, 125
oxidation, 144, 191
oxygen, 4, 6, 71, 72, 155, 194, 195, 231
oxygen consumption, 6

P

pain, 153
palate, 172
palsy, 171, 172
Panax ginseng (Korean ginseng), vii, 1, 2, 7, 8, 9, 23, 44, 45, 55, 81, 83
paprika, 238
paralysis, 11
p-coumarate 3'-hydroxylase (C3H), 105
p-coumaric acid (7), 94, 102, 103
p-coumaroylquinic acid, 94, 97, 103, 105
p-coumaroylquinic acid (14), 94, 103, 105
p-cymene (30), 145
pentacyclic diterpenes, 108
peppers, vii, 214, 217
peritoneal macrophages, 69

peroxisome proliferators-activated receptors-gamma (PPAR-□□) mRNA expression, 13
petroleum ether, 191
p-feruloylquinic acid (12), 94, 103, 105
pH, 114, 116
phenolic compounds, 112
phytochemical compounds, 110
pigments, 217, 231, 234
piles, 171, 172, 174
plant, vii, ix, x, 1, 7, 8, 10, 12, 28, 87, 100, 103, 105, 114, 116, 126, 127, 139, 140, 141, 142, 143, 144, 149, 155, 157, 158, 159, 174, 197, 198, 199, 207, 232, 233
plant metabolism, 105
plants, 152, 154, 193
precursor, 35
progressive damage, 144
promotion, 144
pro-oxidant properties, 144
protection mechanism, 122
protein level, 28, 124
puffed ginseng prepared by puffing at 294 kPa (PG2), 16
puffed ginseng prepared by puffing at 490 kPa (PG3), 16
puffed ginseng prepared by puffing at 98 kPa (PG1), 16
pungent peppers, 230
purine biosynthesis, 106
purine nucleotides, 106
pyrogallol, 220

Q

quantification, 220

R

rabbit, 10
Rana esculenta, 10

Index

rasayana, 176
rate-limiting step, 68, 106
receptors, 6
reduction, 23, 144, 195
reflux, 110, 112, 114, 116
rejuvenative, 176
relationship, 2, 97, 120
reproducibility, 227
resin, 193
resistance, 35
retention factor (Rf), 118
retention times, 118
rheumatism, 174
roasted bean, viii, 93
roasted coffee beans, 110, 111
roasting degree, 112, 114, 117, 122
roasting process, 109, 117, 119
robusta, viii, 99, 125, 127
RT-PCR, 6, 36
Rubiaceae family, 99

S

S type lignin biosynthesis, 105
S-adenosylmethionine (SAM) cycle, 106
safety, 193
Salmonella enterica, 123
saponification, 220
satiety, 121, 130
scanty menstruation, 176
scrofula, 171
scrofulous infections, 172
sedative, 171
seeds, 172, 175, 193
selectivity, 227
semecarpetin (133), 179, 181
Semecarpus anacardium, 175, 181, 184, 188, 189
Semecarpus anacardium L.f., 175, 181, 188, 189
sensorial characteristics, viii, 109

sensory likeable characteristics, 94
separation, 220, 228
Sesamum indicum L., 175
shikimic acid (6), 94
short-term regulation, 121
siderophore-Fe3+ complex, 125
siderophores, 125
silica gel 60 column, 118
silverskin, 99
silybin, 153
skeletal muscles, 120
skin, 171, 172, 176
skin ailments, 176
skin diseases, 171, 172
smooth, 46, 99
Soxhlet diterpenes extraction, 110
spasmolytic properties, 10
sphingosine kinase 1 (SPHK1), 23
sphingosine kinase 1 (SPHK1) activity, 23
sphingosine kinase inhibitor II, 6
sphingosine kinase-1, 2, 86
sprains, 174
St. John's wort, 152
stem, 171
stimulants, 49
stimulatory effects, 97
STPG composite capsule formulation, 6
Strecker degradation, 109
Streptococcus mutans, 123, 124
stress, 5, 73
stress damage, 120
strokes, 97
substrate utilization, 61
sulphate metabolite, 122
supercritical carbon dioxide (SC-CO2) extraction, 110
surface tension, 227
swollen parts, 172
synergic properties, 125
synergism, 122
synthesis, 232
syphilis, 172

Index

T

taste, xi, 144, 212
Terminalia chebula, 175
tertiary butylhydroquinone (TBHQ), 220
tetrahydrorobustaflavone (132), 181
theobromine, 96, 97, 106
theobromine synthase (TS), 106
theophylline, 97
therapeutic potential, 78, 94
thermal degradation, 227
thermogenesis, 123
thin layer chromatography (TLC), 118
thyme oil (24), 145
thymol (25), 145
tissue senescence, 105
tissues, 20, 46
tonic, 171
tooth surface, 124
total cholesterol, 6, 192
tough outer skin, 99
traditional medicine, 174
trans-cinnamic acids, 103
transcription factor, 13
transformation, 192
triglycerides, 6
trigonelline, 94, 97, 101, 103, 118, 123
trigonelline (5), 94, 101, 118, 123
tumor colonies, 21

U

ultrasounds, 110
ultraviolet (UV) light, 118
un-roasted coffee beans, 97
unsaponifiable fraction, 108
urinary output, 176
uterus, 172
utilization, 60, 68
uvula, 172

V

vagus nerve, 11
varnish, 171
vascular cell adhesion molecule-1 (VCAM-1) expression, 120
vascular endothelium, 6, 123
vasomotor center, 11
vata disorders, 176
venereal disease, 171
violaxanthin, 217
virulence, 125
viscosity, 227
volatile compounds, 110
volatile heterocyclic compounds, 103

W

warm blooded animals, 10
weight-loss supplement, 120
whole plant, 152
worms, 171

X

xanthosine (18), 95, 106
xanthoxin, 217

Z

zeaxanthin, 217

α

α-pinene (33), 145

γ

γ-terpinene (27), 145